Harvesting Space for a Greener Earth

Greg Matloff • C Bangs • Les Johnson

Harvesting Space for a Greener Earth

Second Edition

 Springer

Greg Matloff
Brooklyn, NY
USA

C Bangs
Brooklyn, NY
USA

Les Johnson
Madison, AL
USA

ISBN 978-1-4614-9425-6   ISBN 978-1-4614-9426-3 (eBook)
DOI 10.1007/978-1-4614-9426-3
Springer New York Heidelberg Dordrecht London

Library of Congress Control Number: 2014934696

Printed on acid-free paper

Springer is part of Springer Science+Business Media (www.springer.com)

Foreword (From the First Edition)

Gloom and doom sells. That salient point is made by the authors and, for me, is reinforced every time I sit down in front of my favorite cable news channel. Mass shooting at a high school in Iowa. The glaciers are still melting. Terrorists attack a hotel in Pakistan. The whales are going extinct. We hear about loose nukes, lunatics in power, corrupt politicians, child abuse, famines in Africa, tidal waves in Thailand, and so on. A woman in the Middle East is raped; then she is accused of immorality for allowing the attack to happen, and is murdered. Racial conflict shows up in Cambridge, MA. News arrives every day that the United States may sink under its accumulation of debt. And we all know the world is running out of oil.

Despite all this, the authors point out that things have been worse. At the turn of the last century, people in the United States were, on average, not living much past 40. One thinks of the pre-Civil Rights era, of the Depression, of the nuclear confrontation between the superpowers over Cuba. I can recall, during those years, watching workers build an overpass on the Baltimore-Washington Expressway, and wondering why they bothered. At that time, an eventual all-out war seemed inevitable.

Harvesting Space for a Greener Earth has three authors: Greg Matloff, a physicist at the New York City College of Technology; his wife, C Bangs, a Brooklyn artist; and NASA physicist Les Johnson.

Matloff is convinced that science and intelligence will eventually win out over the lunacy that has plagued humankind since Herodotus was writing history. He has published more than a hundred scientific papers, and is the author of, or collaborator on, seven other science books.

Bangs's work might be said to be inspired by starlight. She uses Gaia, goddesses, and the night sky to portray various aspects of the cosmos. Her conversations with her husband about life, death, and the universe provide much of the insight on display in her art. Les Johnson has managed various in-space propulsion technology projects for NASA. He's worked on a tether experiment using Earth's magnetic field for propulsion. And he has twice received NASA's Exceptional Achievement Medal.

In *Harvesting Space for a Greener Earth,* the authors recognize the severity of the problems that humanity faces, and they don't pretend that technological break-throughs alone will be enough to save us. But it is clear that, without the technology, we are headed for a catastrophe.

And there is reason for optimism. After all, life is getting better. We live more comfortably than our grandparents did. It is now possible that African-American neighbors can show up in the neighborhood without unduly alarming those who really like the nineteenth century. We now have the Internet; we are showing signs of getting rid of tobacco, and medical science has come a long way. And we've developed a global sensitivity that we never really had before.

The authors suggest that the growing awareness that we share the same world, and that it has its limitations, began with those first photos of Earth taken from the Moon, the pictures of the fragile blue world drifting through an endless sky. They're probably right. If the human race ever does really coalesce into a family, I suspect those pictures will be hanging near the front door of the family estate.

Will it ever happen? Bangs, Matloff, and Johnson think it will. And, in *Harvesting Space for a Greener Earth,* they lay out a plan to make it possible. If we choose to make the effort, to collaborate, to work together, here, they say, are the tools we will need. Here's how to deal with the inevitable energy shortages that are on the horizon, and do it in a way that does not wreck the ecosystem. They point out that we have a virtually infinite supply of clean energy available, compliments of the Sun. All we need do is make the investment to harness it.

Here also is a technique for getting rid of the pollution caused by various manufacturing industries. And we might also want to take a long look at the dangers presented by near-Earth asteroids. A single rock, a mile or so in diameter, could put the lights out for all of us. Permanently. Most people shrug at the scenario. They would ask how many of them are there? The answer, unfortunately, is that they are numerous. And they are all over the place. Two weeks ago, as I write this, something very large crashed into Jupiter. We never saw it coming.

Then there are the issues of global warming. And conservation of resources. With world population at 6 billion and climbing, recycling aluminum cans, planting more trees, and turning the air conditioner down a notch won't get the job done. Wont' come close. We're face with serious problems, and eventually both will reach the crisis stage.

Do the authors have a plan? You bet! It's not a solution we could manage today, because we don't have the technology yet. But there's time. If we act. If we avoid our usual propensity of waiting until the flood waters are running into the valley. And that's the problem with the plan. It will require political will and advance planning. And maybe, most important of all, imagination.

I'd like very much to see a copy of *Harvesting Space for a Greener Earth* placed in the hands of leaders, and talk show hosts, around the world. The rest of us will be able to profit by it, too.

Jack McDevitt
Nebula-winning author of Time Travelers Never Die

Preface

A lot has happened since the first edition of this volume was published in 2010. First, the order of authors has changed. That's because Greg Matloff elected to take early retirement from his full-time teaching position at New York City College of Technology and Les Johnson's responsibilities at the NASA Marshall Space Flight Center have expanded.

After the publication of the first edition, artist C Bangs created with Greg Matloff's collaboration an artist's book *Biosphere Extension: Solar System Resources for the Earth*, which has been collected by the Brooklyn Museum and the British Interplanetary Society. Some of the art from that volume has been modified to appear as chapter frontispiece art in this edition.

The concept of applying solar system resources to improve terrestrial existence seems a bit more immediate in 2013 because of the increased number of private and public players in the space arena. Diverting Earth-threatening asteroids using our developing interplanetary capabilities has taken on fresh urgency after the 2013 air burst over Siberia that injured about 1,500 people but happily resulted in no fatalities.

Colleagues of ours affiliated with Oak Ridge National Laboratory in Tennessee who authored a chapter in the first edition discussing a space-based approach to alleviating global warming have expanded their efforts to contribute an additional chapter to this second edition on suggested Earth-based approaches to deal with this global issue.

Perhaps the failure of many world governments to adequately address this urgent problem is a cause for pessimism. But such pessimism might be countered by the fact that application of green energy is increasing on our planet.

No one can predict the outcome of the multiple current world crises. But solar system resources can certainly be applied to alleviate some of these problems if we have the will to tap them. It is hoped that the ideas presented in this volume contribute to a positive future for humans and all terrestrial life.

Brooklyn, NY, USA	Greg Matloff
Brooklyn, NY, USA	C Bangs
Madison, AL, USA	Les Johnson

Acknowledgements

We would like to thank Ken Roy, Robert Kennedy, David Fields, and Eric Hughes for their contributions to Chaps. 15 and 16, where they discuss various approaches to mitigate the effects of global warming by geoengineering. We are blessed to have such technically innovative thinkers as both colleagues and friends. We also appreciate the editorial assistance of Maury Solomon and Nora Rawn of the New York Springer office.

Thanks also to Mitzi Adams and Sam Lightfoot for providing their technical expertise in reviewing Chaps. 4 and 6.

We appreciate our colleagues, friends, and students, who have been a constant source of inspiration.

I (Les Johnson) would like to thank Stuart and Dolores Peck for allowing me to use their spare room so that I could have a quiet place to write on Wednesday nights. I also appreciate the constant supply of ice cream they provided. A person could not have better in-laws!

Some of the chapter frontispieces utilize C Bangs's photographs of exhibits in the Hall of Human Evolution at the American Museum of Natural History in New York City.

Contents

About the Authors

Greg Matloff has published or delivered about 100 research papers related to atmospheric physics, space exploration, and space science and has authored or co-authored nine books and many popular articles. Dr. Matloff is a professor emeritus and adjunct professor at New York City College of Technology, CUNY, where he teaches astronomy. Matloff consulted for NASA Marshall Space Flight Center; he is a member of the International Academy of Astronautics, a fellow of the British Interplanetary Society and a Hayden Associate at the American Museum of Natural History, where he worked on asteroid diversion techniques. He also heads the science board on the new Institute for Interstellar Studies.

C Bangs has exhibited her art in museums and galleries throughout the United States, South America, Europe and Australia. Bangs has created chapter frontispiece art for the books authored by her husband, Greg Matloff. Reversing roles with Greg, she created her first artist/scientist book, which has been collected by the Brooklyn Museum and the British Interplanetary Society for their artist book collections. Her work has appeared in the *Journal of the British Interplanetary Society, Analog: Science Fact and Fiction* and *Zenit*. Bangs worked under a grant at NASA Marshall Space Flight Center and then as a NASA faculty fellow for three sequential summers. Bangs' art is included in numerous public and private collections.

Les Johnson is the deputy manager of NASA's Advanced Concepts at the Marshall Space Flight Center in Huntsville, Alabama. Previously he managed NASA's In-Space Propulsion Technology Project, developing advanced technologies such as solar sails and aerocapture for future space science missions. He was the NASA co-investigator on the Japanese T-Rex tether propulsion experiment in 2010. In addition to his NASA credentials, Johnson is also an inventor. He holds three patents and was twice the recipient of NASA's Exceptional Achievement Medal. He is the author of numerous technical publications, co-author of three mass-market popular science books and two science fiction books and has consulted on various novels and two major motion pictures.

Part I
Mythical Paradise

Chapter 1
Introduction: Welcome to the Present

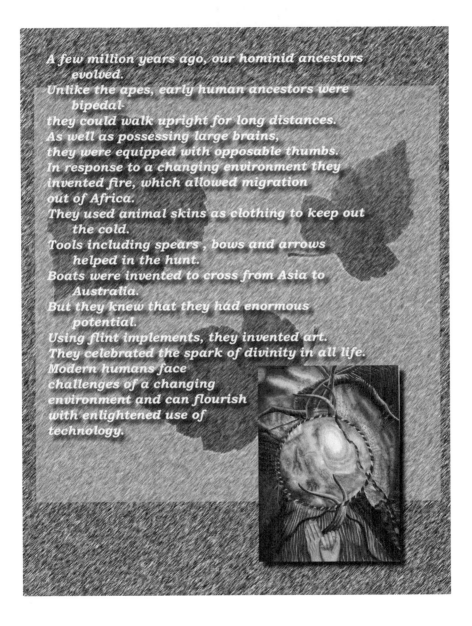

A few million years ago, our hominid ancestors
    evolved.
Unlike the apes, early human ancestors were
    bipedal-
they could walk upright for long distances.
As well as possessing large brains,
they were equipped with opposable thumbs.
In response to a changing environment they
invented fire, which allowed migration
out of Africa.
They used animal skins as clothing to keep out
    the cold.
Tools including spears, bows and arrows
    helped in the hunt.
Boats were invented to cross from Asia to
    Australia.
But they knew that they had enormous
    potential.
Using flint implements, they invented art.
They celebrated the spark of divinity in all life.
Modern humans face
challenges of a changing
environment and can flourish
with enlightened use of
technology.

G. Matloff et al., *Harvesting Space for a Greener Earth*,
DOI 10.1007/978-1-4614-9426-3_1, © Springer Science+Business Media New York 2014

> *"I know a bank where the wild thyme blows,Where oxlips and the nodding violet grows;Quite over-canopied with luscious woodbane,With sweet musk-roses and with eglantine"*— William Shakespeare, from *A Midsummer Night's Dream*

If you are reading this book, you are most likely a member of the most privileged generation in humanity's experience. You have a roof over your head—a vast improvement over the lot of many of our ancestors and a significant number of humans today. You have access to good health-care facilities and can count on between 70 and 80 good years.

Clean water is yours at the twist of a knob. And food—as healthful or exotic as you might desire—is available a short walk, subway ride, or drive from your front door. Your great grandparents might have sacrificed a great deal for these advantages alone.

Education, although not universal, is widespread. In all likelihood, you have a moderately creative and financially rewarding professional life, also a great rarity before about 1950. If you push a button, entertainment and information from the world over can flood your consciousness. You can even remain connected to this planet-wide information network as you walk in a park, cook your dinner, or ride in a car.

However, serious problems loom on our horizon, problems that threaten to swamp our envious situation. The world's population continues to expand. And surprisingly to some, the burgeoning populations of Asia, Africa,, and South America desire the same advantages enjoyed for many years by North Americans, Europeans, the Japanese, and Australians.

Can contemporary civilization provide for billions of more humans living as well as we do? Where will the energy come from? How do we deal with the pollution? Will carbon dioxide emissions and other human-produced greenhouse gases evoke long-lasting climate change that will increase global temperatures, raise sea levels, and swamp coastal lands?

A million years ago, as our ancestors began to emerge in the park-like savannah of central Africa, a response developed to the problem of environmental degradation. When your local environment was exhausted, move! This worked well as hunter-gatherers moved around the globe and civilizations developed and spread. The world is littered with the ruins of once-great cities surrounded by degraded environments.

However, today there is nowhere to flee. Civilized humans are everywhere on Earth. Early science-fiction authors hoped for benign climates on neighboring worlds, but none of these worlds could sustain more than a tiny fraction of the human population, and that at great expense. Even if people were genetically modified with gills to live in the oceans, this would be a mere stopgap measure; the oceans are not immune to human pollution.

Conservation, limiting growth, and recycling can provide some relief, but any such relief will almost certainly be only temporary. With the rest of the world's economies growing into variations of our capitalistic one, it is only a matter of time before we simply run out of resources, energy, and places to store our waste. This does not mean we should not conserve, limit growth, and recycle! On the contrary, these measures are essential to the survival of our civilization and, potentially, our species. They are simply insufficient to resolve the core issue we face—that Earth cannot by itself indefinitely sustain a worldwide population of consumers. It is impossible to recycle with 100 % efficiency; to limit growth to "no growth," or to conserve into prosperity.

It was a bitter cold day in December 1968 when humans in general became aware of their kinship as riders on a fragile, living Earth. From a quarter-million miles away, the crew of *Apollo 8* pointed their cameras homeward after they had safely settled into humanity's first orbit of the Moon.

The view of the desolate Moon was striking on our television screens, and the astronauts' scripted reading from Genesis was stirring. But it was the shimmering, living, marble-sized Earth, hanging in stark contrast above the stark lunar horizon, that would profoundly alter our view of ourselves and our world.

Living worlds are fragile and delicate. And space is very, very large. This lesson would be repeated and amplified in 1990 when on the edge of the galactic void, the cameras of *Voyager 1* were focused back on our planet. From its multi-billion-mile vantage point, *Voyager* imaged Earth as a pale blue dot almost lost in the glare of the distant Sun. Earth is a seemingly unique abode for life in an otherwise empty and apparently lifeless expanse of nothingness.

So as you escape our increasingly urbanized world to stroll through your local park, botanical garden, or forest and gain spiritual sustenance from your temporary immersion in this sanitized (predator-free) version of our original environment, you might well ponder the troubled legacy of our golden age. Planets abound in our galaxy, but planetary ecospheres are rare and precious.

Will the parks of Earth survive this age of over-population, resource consumption, nuclear proliferation and terrorism? Or will our civilization go down with a bang or a whimper, to be remembered in legend by a remnant population eking out a living in a depleted, contaminated landscape?

No one can know the answer to this question. But there is hope within the gloom. An optimistic scenario exists if contemporary humans are collectively wise enough to grasp the opportunities of the present. Although the modern environmental movement applies much-needed first aid to our resource and environmental challenges, we can simply look up for the potential cure—space harbors enough resources to meet the needs of an ever-more prosperous humanity for millennia. Space is an environment generally hostile to life that can be used to house industries whose byproducts are also antithetical to life, and nearly infinite energy.

Our Solar System is very rich in energy and resources and can even serve as a sink for some of the unavoidable effluents of technological society. Further, to avoid the fate of the dinosaurs, humans may decide to use our revived interplanetary capabilities to alter the solar orbits of those kilometer-sized chunks of cosmic rock and ice that occasionally wallop Earth. And if we can move these objects around, perhaps we can mine some of them.

Even if interplanetary space can never absorb more than a handful of Earth's living inhabitants, its resources can be used for the betterment of life. Envision a future where plentiful energy comes from the Sun, industrial pollution is virtually removed from the ecosphere, and no country needs to suffer from a lack of natural resources. Then, the 10- to 15-billion peak human population on Planet Earth can enjoy comfortable, productive lives. And the parks of Earth, the thyme and violets, need not die.

This book is divided into three parts. Part I, Mythical Paradise, reviews current scientific thinking about how Earth came to exist and how life arose. Many consider primitive Earth, the planet as it was before the rise of human civilization, to be a paradise lost. In reality, it was a hostile environment in which only the strongest survived, and it was far from being a paradise.

Part II, Paradise Lost?, describes the rise of human civilization, our progress from simple, daily survival to where we are today—with many humans living long, productive, and meaningful lives—and the associated (mostly negative) impacts to the environment that our civilization has wrought. It is here that we outline the challenges our twenty-first century civilization faces.

Part III, Sky Harvest, describes how space and space technologies can be used to monitor the global environment, help undo ecological damage, and prevent further damage to Earth's ecology upon which we all depend.

The first edition of this book was published in 2010 under the title *Paradise Regained: The Regreening of Earth*. In a few short years, a lot has happened that affects the issues discussed in this book.

On one hand, a lingering global economic crisis and unending regional wars have led to widespread doubt regarding the ability of national leaders and the international community to successfully address the complex issues facing us. This, in turn, has led to widespread pessimism about our future prospects.

On the other hand, we have witnessed the rise of commercial space enterprises. Traveling in space, exploring this vast domain and ultimately exploiting the resources that abound there need not be the exclusive domain of governments and international consortiums.

The survival and advancement of our now-global civilization presents us with major challenges. Nevertheless, we should try to preserve optimism and press ahead with the job. Above all else, we should not lose heart! If this book helps to restore hope for an expansive, prosperous future for humanity, it has succeeded.

Chapter 2
Space Utilization: A Moral Imperative

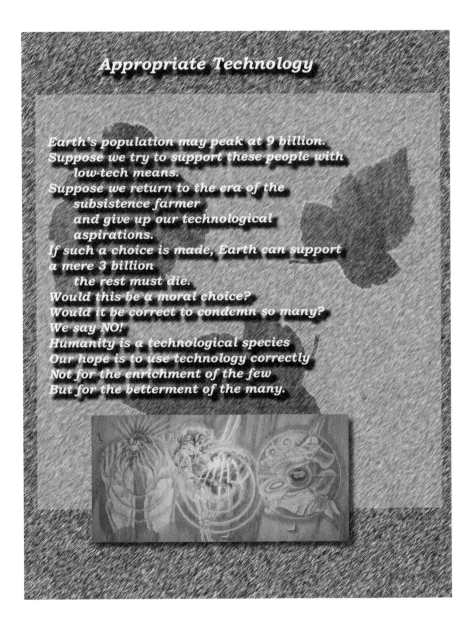

Appropriate Technology

Earth's population may peak at 9 billion.
Suppose we try to support these people with
   low-tech means.
Suppose we return to the era of the
      subsistence farmer
      and give up our technological
      aspirations.
If such a choice is made, Earth can support
a mere 3 billion
      the rest must die.
Would this be a moral choice?
Would it be correct to condemn so many?
We say NO!
Humanity is a technological species
Our hope is to use technology correctly
Not for the enrichment of the few
But for the betterment of the many.

G. Matloff et al., *Harvesting Space for a Greener Earth*,
DOI 10.1007/978-1-4614-9426-3_2, © Springer Science+Business Media New York 2014

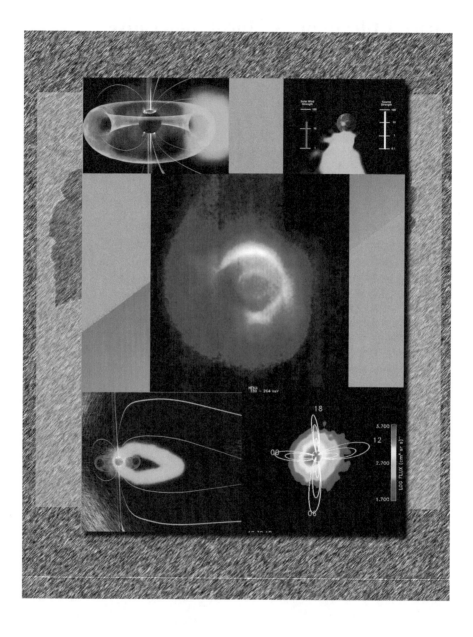

"I saw the sudden sky
Cities in crumbling sand;
The stars fall wheeling by;
The lion roaring stand."
—From the poem "The Lion" by W. J. Turner

As citizens of the already-developed countries of the world grapple with their impact on the planet, so do those living in the rising economies of China, India, and Africa. But although it is relatively easy for those of us in the developed world to say we will make due with less consumption for the good of the planet, those in the developing world don't have that luxury. They need to consume more resources to have jobs, educate and feed their children, and to generally improve their quality of life. On a per capita basis, they don't want or need to be as profligate in their use of the world's resources as we've been, but there are so many of them that each using only using a small fraction (as compared to residents of North America, Europe, and Japan) of what their developed counterparts use will place dramatically increased stress on the global environment and the planet's resources.

Space advocates and environmentalists are cut from the same cloth. They have very similar long-term goals and the same basic moral principles guiding them. Life on Earth is precious; it must be preserved and protected for future generations of the planet's inhabitants, both human and nonhuman. Unfortunately, as we look at the challenges posed by rampant materialism, increased production (using up the planet's resources), and the commensurate increased pollution from the developing world, these two groups tend to take very different approaches to addressing these core problems. Worse yet, each group tends to view the other with suspicion and questions their motives and tactics. Space advocates tend to believe the environmentalists are motived by a political and economic agenda (tending toward the political left), while environmentalists tend to believe that space advocates are part of the military/industrial complex following a right-wing political agenda. Both are simplistic generalizations. If the two groups are serious about saving the planet then they need to put aside their political differences and work together. Environmentalists are working to stop very real and immediate environmental damage, but they don't have a long-term plan for saving the planet. Space advocates have the long-term solution to environmental stresses on the planet—the use of space resources—but tend to ignore what can be done in the short term, to buy time until the long-term solution can be implemented.

It is this long-term solution that motives the authors to write this book. We mirror the divide described above, with at least one of us leaning politically left and the other politically right. Yet we agree that must act now and plan for the future, if life on Earth is to have a chance. And we must work together on a just and moral foundation.

Since there have been people around to consider the notion philosophers and theologians have debated what constitutes morality. The word *morality* comes from Latin and refers to our notion of what constitutes right and wrong or good and evil. What constitutes a moral action varies dramatically from culture to culture, though many cultures share some common thought as to what is right and wrong. For example, most modern cultures believe that murder is not a morally acceptable method to settle a dispute and that theft is not a moral act. But even these seemingly simple moral judgments are far from universal, and anthropologists can undoubtedly find cultures on the globe that do not share these views.

We will put forth a moral concept upon which this book will stand. It is a concept that should have nearly universal appeal and should guide much of our decision making regarding both space and environmental policies. The moral statement that drives our thesis is deceptively simple. In fact, it is so simple that an entire company is built around it as both a trademark and a pithy statement: Life is Good™. The converse is also a moral assertion—that which leads to non-life is evil. And by "life" we mean communal life, not that of an individual animal or plant, although in the case of humans we believe the assertion is valid almost 100 % of the time. Those who seek to preserve life, human and nonhuman, are acting in a morally superior manner compared to those who seek to diminish or harm life. Often we must act on this moral principle in such a way that some life is harmed along the way to a more global solution that greatly preserves or improves the quality of life in general. These choices must be carefully considered in the context of whether or not they will improve the quality of life or maintain life in the "big picture" or on a scale beyond the immediate and obvious impacts to an individual.

We believe the moral decision that life on Earth is good drives those in the modern environmental movement to their activism. It is this same moral decision that motivates many space enthusiasts, activists, and professionals.

It is a moral declaration to say that life is good and better than non-life. Once this declaration is made, it is easy to see that in order for humans to prosper and be good stewards of the planet and its myriad life forms, they must stop destroying the environment that gives it life. But how do we accomplish this goal and maintain all of the positive aspects that come from our twenty-first-century technological civilization? The answer is pretty simple. We must place heavy industry, with all of its inevitable pollution, in a place that is already hostile to life and in which life will almost certainly never arise—in space or on the Moon. A space-based infrastructure is not only possible but also essential if we are to restore our home planet and turn it into a safe place to live, pollution free. This chapter explores the current environment of near-Earth space, so as to address concerns by some that our presence there will somehow pollute it.

Most of us think of outer space as being empty. If we are comparing space starting at about 200 km above our heads and extending far outward, to what it is like around us on Earth, then *empty* seems to be the appropriate word. For all practical purposes, to most humans, space is empty. (When we refer to space in this context, we are referring to that which is separate and apart from a planet, comet, or asteroid. Of course, the total volume of space occupied by planets, comets, and asteroids is so small in comparison to the rest that it is inconsequential overall.) In space, there is no atmosphere to breathe, no potable water to drink, and not much else in significant quantity—at least from the average human being's perspective.

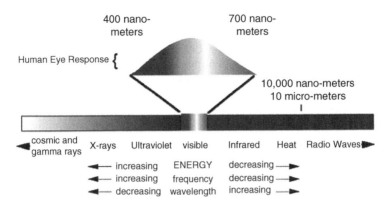

Fig. 2.1 The Sun emits light at many wavelengths, only a small fraction of which can be seen by the human eye. This portion of the spectrum is known as "visible light" (courtesy of the Universities Space Research Association)

However, in the 50 years of space exploration, scientists have learned that space is far from empty, and most of what is found in space is directly antithetical to life. To begin with, the entire Solar System is bathed in sunlight. We depend on the light emitted by the Sun to sustain us. But this light is only a small part of the total electromagnetic radiation ("light") emitted by the Sun. We see only this part because our dense atmosphere filters out most of the rest. The Sun emits light at several wavelengths, from infrared to extreme ultraviolet. The part of the Sun's spectrum that we see is only a very small part of the emitted spectrum and is aptly called "visible light" (Fig. 2.1). By definition, the other parts of the spectrum are invisible to our eyes, but not to our scientific instruments.

Over the past several years, there has been much discussion about solar ultraviolet light. It is this part of the spectrum that causes people to tan and sunburn—in many cases, subsequently causing skin cancers. Fortunately for us, most of the ultraviolet light from the Sun is filtered at high altitudes by atmospheric ozone. There are multiple wavelengths of light that are considered to be part of the ultraviolet spectrum. They are divided into two types: ultraviolet A [UVA] and ultraviolet B [UVB]. The amount of UVB light that penetrates through the atmospheric ozone decreases rapidly with increased ozone concentrations. The converse, unfortunately, also applies: decreased ozone will dramatically increase the amount of UVB reaching the surface of Earth. UVA passes roughly unhindered through the atmosphere.

Since the 1980s, and mostly likely beginning earlier, a global decrease in atmospheric ozone density was observed, with especially large depletions occurring near Earth's poles. The large depleted regions near the poles were dubbed "ozone holes," and their existence was blamed on the human-caused emissions of ozone-depleting chemicals such as chlorofluorocarbons. This depletion was of concern because increased UVB exposure would increase the number of people getting various cancers. It is also thought that increased UVB would do significant harm to plankton in the oceans, with damaging ripple effects felt throughout the food chain that depends on plankton. Also, several crop species are thought to be UVB sensitive, with death or significantly reduced yields resulting from increased exposure. Global environmental action was taken in the 1990s to reduce human emission of ozone-depleting chemicals so as to mitigate further ozone depletion.

Solar ultraviolet radiation acts as a sterilizer and quickly kills any unprotected life. In fact, ultraviolet sterilization is used commercially to kill bacteria in swimming pools and in air purification systems. In space, there is no atmospheric ozone to filter any of the UVB, and ultraviolet radiation from the Sun is deadly.

In addition to visible and ultraviolet light, the Sun emits many other forms of radiation. High-energy electrons and protons continuously stream from the Sun; this stream is commonly called the solar wind. The Sun emits these particles with velocities between 200 and 800 km/s, depending on where you are relative to the Sun's equator and the solar activity cycle. When these charged particles encounter matter, such as living tissue, they deposit their energy as they slow down and stop therein. These particles have a lot of energy, and the slowing-down process does significant damage to any living tissue in which it occurs. Even short-term exposure to unfiltered sunlight can cause mutations and cancer. Moderate to long-term exposure results in significant cell damage and death.

And it gets worse. In addition to this somewhat steady stream of charged particles coming from the Sun, there are periodic bursts of intense high-energy radiation, called solar energetic particle events, that blast lethal storms of charged particles outward from the Sun to the outermost regions of the Solar System, each packing enough energy to cause human death within minutes to hours of exposure.

Fortunately, our Earth once again protects us from this danger. Earth's relatively strong magnetic field acts as a radiation shield, deflecting all but the most energetic of these particles away from the surface of the planet. (Charged particles, when moving through a magnetic field, experience a force acting on them, in this case, a deflecting force.) Our thick atmosphere, which has the approximate stopping power of 10 m of water or 4 m of concrete, absorbs most of the remainder (Fig. 2.2).

On Earth, we go blithely through our days during these storms, blissfully unaware of the hellish inferno blasting through space a mere few hundred kilometers above our heads.

Across the globe, the average temperature does not vary by much, providing a reasonably stable thermal environment for all sorts of life. Where seasonal variation does occur, life, in general, has adapted to it, growing during the warmer summer months, hibernating during the colder winter months. In author Les Johnson's adopted hometown of Madison, Alabama, the average temperatures range from a low of about 29° Fahrenheit in January to a high of 89° in July. Most residents of north Alabama adapt quite nicely to this range of 60°. Thanks to the complex biosphere of Earth, with enormous oceans and a thick atmosphere to average out and regulate thermal effects, 60° temperature variations in the course of a year are typical for Earth's nontropical, nonpolar regions, which are usually mild and for which life has readily adapted.

In space near Earth, without an atmosphere or ocean, the temperature can vary from +200 to −200°F as quickly as an object moves from being in sunshine to being

Fig. 2.2 Earth's magnetosphere, depicted here, acts as a shield against all but the most energetic cosmic rays, preventing them from reaching the surface of Earth in great quantities (courtesy of NASA)

in darkness. Most forms of life would either freeze to death or be cooked in these 400° temperature swings.

If you go out on a clear night in the middle of August, you have a pretty good chance of seeing a shooting star, more accurately called a meteor. If you know where and when to look, you will likely see several meteors. It is during this time that the Perseid meteor shower occurs, a result of Earth's passing through the trail of pea-sized debris left behind by comet Swift-Tuttle. Harmless to us on the surface of the planet, these small pieces of cosmic dust and ice hit the atmosphere at 60 km/s and quickly burn up, leaving a glowing trail across the night sky. The Perseids are a harmless, beautiful sight, unless you happen to be above the protection of our dense atmosphere.

If you are in space or on the Moon, these pebbles traveling at 60 km/s are more deadly than a speeding bullet (which travels at only about 0.8 km/s). When being hit by a bullet, a fist, or a car, what matters is not the overall speed but the kinetic energy of the impactor. For the most part, the energy transfer from the impactor to you determines how much damage you sustain. Kinetic energy is the energy associated with the motion of a body, and the amount of kinetic energy depends on both the mass of the object and the square of its velocity. In practical terms, this means that you can double the amount of damage done by doubling the mass of the projectile and simply keeping its speed constant. However, if you double its speed, you do not double its kinetic energy; you increase it by a factor of 4. A bullet shot from a gun on Earth will have only a fraction (about 0.017 %) of the energy of that same bullet moving with the speed of a Perseid meteor. It will therefore do only a fraction of the damage caused by being at the wrong place at the wrong time and getting hit by a meteor!

Granted, the probability of getting hit at any given time is low. There are only so many meteors, and space is very big. But once you factor in the frequency and damage potential of the larger impactors, such as asteroids, the risk begins to rise sharply. Getting hit with a large rock moving at tens of kilometers per second is like having a bomb dropped on you, or worse. (See Chap. 14, in which we discuss the threat from near-Earth objects and how one object impacting Earth is thought to have ended the reign of the dinosaurs.) The bottom line is that we live in a cosmic shooting gallery, and getting hit by any of these objects would be the perfect definition of a bad day.

One aspect of space that is foremost in most people's minds when they think of life in space is vacuum. What happens to living things in the vacuum of space? This is an easy question to answer. Unlike depictions in many science-fiction movies, the human body does not inflate like a balloon and explode when exposed to vacuum, nor does the blood boil or immediately freeze. You may not even immediately lose consciousness. But unless you get back under pressure very quickly, you will most assuredly die—from asphyxia.

If you were to step out on the surface of the Moon and somehow survive the vacuum, you would still have to worry about the continuous exposure to solar radiation in all its forms, such as visible light, ultraviolet light, and infrared light, as well as the charged particles in the solar wind and those resulting from solar energetic particle events. This radiation can quickly reach levels that even a moderately protected human could not survive. Next, you would have to protect yourself from the extreme heat during the 14-day lunar day and the extreme cold of the 14-day lunar night. (The Moon rotates more slowly on its axis than does Earth. One lunar day is therefore equal in length to 14 Earth days. We see only one side of the Moon because its rotation rate corresponds to the rate of lunar revolution about Earth.)

If you are standing on the Moon, you are standing on what may be the deadest place in the nearby Solar System. The Moon is blasted by intense solar radiation, remains in vacuum, and is alternatingly cooked and frozen on a regular basis. There is probably not a deader environment anywhere close by in the Solar System and certainly not anywhere on Earth (Fig. 2.3).

Without human intervention, space is anti-life. With the possible exception of the bacterium *Deinococcus radiodurans*, no life as we know it can survive in this hostile environment without the artificial protections such as those that we humans devise for our space explorers. Given that there is no air or water to pollute, no ozone to deplete, no climate to change, no environment to harm with the radioactive by-products of our nuclear power plants (the radiation from the Sun makes our pitiful human radioactive output seem truly minuscule by comparison), and no

Fig. 2.3 The Moon is a dead environment, without any life—until we go there and build our cities and factories (courtesy of NASA)

ecosystem to clutter with our waste products, what can we humans possibly do in space to make it more anti-life than it already is? Absolutely nothing.

In fact, the argument can be made that by expanding the realm of human activity to space, including all the processes and products that on Earth would be called pollution and pollutants, we will be creating new places for life to exist and thrive. Such expansion would be a thoroughly positive moral choice. Our industrial plants will have to have breathable air and drinkable water, they will have to have artificial protection from solar radiation in all its forms, and they will have to regulate the temperature so that human life can survive and thrive. We will be creating "green" ecosystems from desert, and the inevitable by-products of our civilization, the pollutants, will not harm any ecosystem in any way. We should not be profligate and wasteful by any means. Our explorers and industrialists will not want to waste anything that has potential use, because it simply will be too expensive to replace. Recycling should be the norm and only after all other options are exhausted should we discard our waste into the space environment.

It may now be difficult to believe, but before 1992, the only planets we knew existed were in our own Solar System. Since then, about a thousand planets circling other stars have been found. Given that there are well over 250 billion stars in our

Fig. 2.4 So far, the only extrasolar planets found are within a very small part of the Milky Way Galaxy (image courtesy of NASA/JPL)

Milky Way Galaxy, and that all we've found so far are within only a few light year of Earth, we can estimate that there are hundreds of billions of planets yet to be found in our galaxy alone. Figure 2.4 shows the region of space within which all of the 'found' planets reside. As you can see, we have a lot more space to explore!

As our technology advances,[1] there is little doubt we will find more planets and be able to better understand the ones we find. Perhaps there some that could support

---

[1] There are various techniques used to find planets circling other stars. The most common is the *Doppler method*. A star's gravity tugs on its planets, keeping them in orbit around the star. But the planet, also having mass, tugs also on the star. The star is far more massive, so the effect of the planet pulling on it is quite small. But not so small that it cannot be detected. As the planet orbits the star, the star also moves in its own orbit around the center of mass of the star/planet system. Using telescopes, we can detect the resulting variations in the star's velocity as it moves either toward or away from Earth as a result of the planet tugging on it by measuring the Doppler shift of the starlight. Just think of the sound a train makes as it is approaching you and then as it moves away. The pitch, or frequency of the sound, changes depending upon its motion relative to where you are standing. This also happens with light, and we have telescopes sensitive enough to measure it.

Another way to find extrasolar planets is by measuring the decrease in brightness of a star that results when a planet crosses in front of it along the line-of-sight between the star and Earth. When a planet crosses in front of its parent star, the observed brightness of the star decreases temporarily.

human life, but let's not bet on them being found anytime soon. The ones found so far don't appear to be hospitable to life as we know it. And, even if we do find other 'Earths' out there, we are nowhere near having the capability to visit them.

Since we don't know of any other place where life exists, let alone life as we know and depend upon here on Earth, we should be doing everything in our power to protect and preserve it for all of its inhabitants. There is only one environment that matters, and we should not hesitate to use space and space resources to preserve it.

We have a moral obligation to develop space resources and to foster space industrialization. To not do so is ultimately anti-life and an immoral act of omission.

From measuring the intervals in which this occurs and the amount of dimming, we can estimate both the size and the star-to-planet distance. This is called the *transit method*.

There are other ways to detect extrasolar planets. An excellent reference is The Laboratory for Atmospheric and Space Physics at The University of Colorado at Boulder, http://lasp.colorado.edu/education/outerplanets/exoplanets.php#detection.

Chapter 3
The Formation of Earth and the Solar System

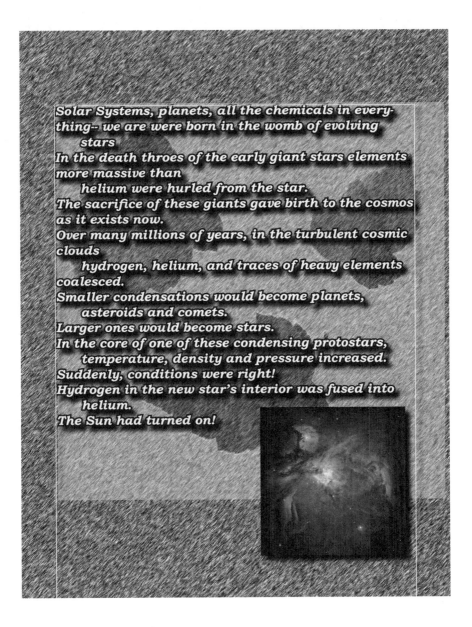

Solar Systems, planets, all the chemicals in every-
thing-- we are were born in the womb of evolving
    stars
In the death throes of the early giant stars elements
more massive than
    helium were hurled from the star.
The sacrifice of these giants gave birth to the cosmos
as it exists now.
Over many millions of years, in the turbulent cosmic
clouds
    hydrogen, helium, and traces of heavy elements
coalesced.
Smaller condensations would become planets,
    asteroids and comets.
Larger ones would become stars.
In the core of one of these condensing protostars,
    temperature, density and pressure increased.
Suddenly, conditions were right!
Hydrogen in the new star's interior was fused into
    helium.
The Sun had turned on!

G. Matloff et al., *Harvesting Space for a Greener Earth*,
DOI 10.1007/978-1-4614-9426-3_3, © Springer Science+Business Media New York 2014

"It lies in Heaven, across the flood
Of ether, as a bridge
Beneath the tides of day and night
With flame and darkness ridge
The void, as low as where this earth
Spins like a fretful midge"
                    —From the poem "The Blessed Damozel" by Dante
                    Gabriel Rossetti.

In the beginning, a cloud composed mainly of hydrogen and helium gas drifted through the interstellar void. Near or within this immense nebula (it must have been trillions of miles across), a bright star blazed. Much larger and more massive than our present-day Sun, this nameless star approached the end of its life cycle about 5 billion years ago. As its nuclear fires waned, it began to collapse. As it collapsed, temperature and density near its core increased dramatically.

Suddenly, conditions in the stricken star's interior became sufficient to support nuclear reactions more elaborate than the hydrogen fusion cycles that support our Sun and most other mature stars. In a spectacular explosion called a supernova, elements as massive as uranium were bred.

For a short while, the dying star outshone the entire stellar host in our Milky Way galaxy combined. Before its embers faded to obscurity, the heavy-element-laden gases it emitted began to mix with the lighter elements in the neighboring nebula.

As remnants of the exploded star diffused through the giant nebula, turbulence developed. In slow motion (at least from the human viewpoint), eddies and whirlpools began to develop throughout the vast celestial cloud. The largest of the eddies, condensing under self-gravitation, would eventually become stars. Smaller eddies would someday result in planets, icy comets, and rocky asteroids.

The nebula had now become a star nursery. Through the telescopes of any advanced extraterrestrial civilization in that distant era, the nebula containing the infant Sun and Solar System (and many others) may have resembled Messier 42,[1] the great nebula visible through binoculars below Orion's Belt in the northern skies of spring.

Eons passed. Material gathered in the large proto-star eddies and in the smaller proto-planets. Continent-sized fragments of rock and ice careened through the infant solar systems, occasionally colliding with a proto-planet or being gobbled up by a proto-star.

Although little has survived on Earth from this ancient era, asteroids and comets are remnants of the Solar System's formative phase. And our telescopes on Earth and in space offer tantalizing glimpses of star formation still occurring throughout our Milky Way Galaxy.

---

[1] The French astronomer Charles Messier in the 1970s cataloged numerous objects in the night sky that were not planets, stars, or comets with a series of numbers beginning with the capital letter M. He did not know what these objects were. The Orion Nebula is an interstellar cloud of dust, hydrogen gas, and plasma from which new stars are born.

As matter accumulated around the proto-Sun, the temperature and density near this object's interior gradually increased. Suddenly, these were sufficient to support a thermonuclear reaction dubbed "hydrogen burning."

A hydrogen-burning star, such as our Sun, emits light through this process. Deep within the stellar interior, hydrogen nuclei are fused together to produce helium, energy, and mysterious neutrinos.

The neutrinos, which have little or no mass, are particles that are non-reactive with normal matter. They rapidly traverse the star's interior layers and disappear in the depths of space, carrying with them a significant fraction of the fusion reaction's energy output. Most of the energy released in the stellar interior is a type of non-visible, high-energy light called gamma rays. By the time this energy has reached the star's outer layer (called the photosphere), much of it is in the form of light visible to our eyes.

After millions of years of contraction, the Sun had finally turned on! First, the pressure of the solar radiation slowed and stopped the contraction. Light from the infant Sun streamed through the turbulent new Solar System. The stream of magnetically accelerated high-energy electrically charged particles called the "solar wind" started up and soon began to influence the nebula regions closest to the infant star.

Near the Central Fire

About 4.7 billion years pass. The planets begin to take form. Although the planets initially are mostly hydrogen and helium, as is the Sun, and the solar wind drives off the initial cloud of light gas enshrouding each of the inner planets.

Close to the Sun, tiny moon-like Mercury coalesces from the dust, gas, and rock in the inner Solar System. With too little mass to long maintain an atmosphere or oceans and much too hot, this tiny world will forever remain barren. But comets in large numbers traverse the young Solar System. Some of these impact Mercury, apparently leaving frozen water deposits in Sun-shielded craters near the planet's poles.

Farther from the Sun is the planet Venus. Since this namesake of the ancient love goddess is a near twin of Earth in terms of size and mass, early astronomers hoped that swamps or oceans might exist beneath the cloud banks of this world. But alas, probes from Earth have penetrated the clouds to learn of this planet's true conditions. Temperatures at the surface are hot enough to melt lead, and atmospheric pressure at the surface of Venus is about 90 times that at Earth's surface. In addition, the atmosphere of Venus is mostly carbon dioxide—a waste product of terrestrial animal life. Even worse, a steady rain of highly corrosive sulfuric acid drips from the leaden skies.

Some love goddess! Even the best-shielded probes from America and Russia have survived no more than a few hours on the surface of this world. Although it's a lovely sight in the evening or pre-dawn sky, Venus is about as close to the medieval concept of hell as any world we are likely to discover. It is unlikely that humans will ever visit this hothouse world, let alone live upon it.

A number of theories have been proposed to explain what went wrong on Venus. One likely possibility is the fact that comets and asteroids near Venus are closer to the Sun than they are when in Earth's vicinity, and consequently move faster. When such objects impact Venus, their high kinetic energy relative to that planet is converted into heat.

Combined with the higher ambient temperatures on this world (since its separation from the Sun is about 70 % of the Earth-Sun separation), conditions were such that the oceans brought by impacting comets evaporated almost immediately. Venus was shrouded from the start by an atmospheric envelope of carbon dioxide. This gas absorbs infrared radiation re-emitted by the planet, causing the "runaway greenhouse effect" that has precluded the evolution of any life we could imagine beneath the cloudy veil of the love goddess.

The next planet out, about 93 million miles from the Sun, is Earth. We'll return to Earth after we survey the formation of other Solar System bodies in the early Solar System.

About 50 % further from the Sun is the Red Planet, Mars. Perhaps because of its red color, this world is named after the ancient war god. Although Mars has a day only slightly longer than that of Earth, it is only about 10 % as massive as Earth. Like Earth, Mars was bombarded early in its existence by water-bearing comets. It also has volcanic mountains larger than any on Earth. In spite of these sources of volatiles and its distance from the Sun, Mars' atmosphere is far thinner than that of Earth. The planet's small size probably caused most of its atmosphere to escape into space.

Unlike Venus, Mars still has water reserves. There is some water vapor in the thin atmosphere and frozen in the polar caps. Recent observations by Mars-orbiting probes and rovers indicate that some water may also exist in the upper soil layers of the Red Planet. Although Mars may have developed life early in its history, living Martians appear to be very sparse or absent.

The Outer Realm

Although chunks of rock and ice routinely traversed the early Solar System, collisions with planets and satellites has gradually cleared most interplanetary space. Beyond Mars, we find the Asteroid Belt. Even though some mountain-sized space boulders are found elsewhere in the Solar System, most of the surviving rocky debris from the Solar System's origin is now found in this realm. Some asteroids are as big as the state of Texas and are classified as dwarf planets.

Sometimes, because of collisions or orbital perturbations, objects from this Asteroid Belt reach Earth. If they survive the fiery descent through Earth's atmosphere, they are called meteorites. Meteorites come in three basic classes—rocky, stony, and carbonaceous. Most rocky and stony meteorites originated in the Asteroid Belt. Water-rich carbonaceous meteorites may have their origin further afield among the comets.

Beginning about five times Earth's distance from the Sun is the realm of the giant worlds: Jupiter, Saturn, Uranus, and Neptune. These worlds are rich in gases, including hydrogen and helium, probably because sunlight and the solar wind are too weak at these solar distances to have evaporated them into space.

The largest of them, Jupiter, has about 318 times as much mass as our Earth, but only about 1/1,000 the mass of the Sun. Composed mostly of hydrogen and helium, Jupiter is often referred to as a "star that failed." If this giant world were somewhat more massive, thermonuclear fusion would have ignited in its interior. We would then live in a double-star system!

But even as a planet, Jupiter is pretty impressive. It is equipped with colorful cloud bands and an atmospheric disturbance larger than Earth called the "Great Red Spot." Like the four largest of its many satellites (Callisto, Ganymede, Europa, and Io), these features are easily viewed through binoculars or a small telescope. These large moons of Jupiter are worlds in their own right. Io features many active volcanoes; a tantalizing, mostly frozen deep-water ocean covers Europa. Many biologists expect to find life in the liquid regions of this satellite-wide sea. Although the largest four Jovian satellites likely coalesced in position like a miniature Solar System during our planetary system's dawn time, most of the smaller satellites of this huge world are captured asteroids or comets.

As is true for all of the giant worlds, Jupiter is equipped with an encircling ring. Planetary rings originate either from the disintegration of satellites that approach the giant world too closely or from debris from a satellite that could not form due to the proximity of the planet. After the Sun, Jupiter is the most intense natural radio-frequency source in the Solar System. If you tune a short-wave radio to 20.10 MHz, much of the static you hear originated from Jupiter! Jupiter has a strong magnetic field and radiation belts that might pose a problem to future human visitors.

The average separation between Jupiter and the Sun is about five times the Earth–Sun separation. Almost twice as far from the Sun is the next major planet, Saturn. In terms of mass, Saturn would seem outclassed by Jupiter, since it's less than a third the mass of that world. Like Jupiter, Saturn has many satellites. One of them, Titan, is the only planetary satellite in our Solar System with a dense atmosphere. Although present-day terrestrial life would have a hard time breathing Titan's nitrogen-argon-methane atmosphere or swimming in its hydrocarbon seas, conditions on this fascinating world may be similar to those on the primitive Earth (although a lot colder).

Through binoculars or a small telescope, Saturn is an elegant sight. Its dramatic rings are spectacular under most viewing conditions.

Further out from the Sun, in the frigid wastes of deep space, are the final two gas giants of our Solar System—Uranus and Neptune. Methane and ammonia are major atmospheric constituents of these worlds, which are respectively about 15 times and 17 times the mass of Earth.

The rotation axis of most planets is usually fairly close to perpendicular to the direction of the planet's solar orbit. Not so for Uranus! Early in the Solar System's history, something huge must have smacked this world. Its rotational axis is nearly parallel to the direction of its solar orbit. This cosmic train wreck would have been something to behold!

Beyond the Planets

Beginning at the orbit of Neptune (about 30 times farther from the Sun than is Earth) and continuing outward about the same distance, we encounter the icy objects of the Kuiper Belt. Dubbed KBOs, these dwarf worlds are composed mostly of frozen water, ammonia, and methane. When a collision or gravitational perturbation pushes a KBO sunward, some of this icy material melts. Heated by sunlight and affected by the solar wind, this material streams from the tiny, rocky nucleus of the KBO. It appears in our sky during these sunward passes as a short-period comet (meaning that its solar orbital period is less than 200 years).

The second largest known KBO, Pluto, was originally classified as the Solar System's ninth major planet. Since the discovery of Eris, a KBO more massive than Pluto, the largest KBOs are now classified as dwarf planets. Most known KBOs are about 100 km in radius. A few, like Pluto, are about 1,000 km in radius. Assuming the existence of many smaller objects, there may be as many as 70,000 members of the Kuiper Belt.

Beyond the Kuiper Belt is the Oort Cloud, a much larger comet repository that originated with the Solar System. Perhaps a trillion comets, each more than 10 km in radius, inhabits this annulus, which stretches to perhaps 100,000 times the Earth-Sun separation. Sometimes, a passing star nudges one or more of these frigid worldlets sunward. It then appears briefly in our skies as a long-period comet. Some of these have orbital periods of 100,000 years or longer.

Meanwhile, Back on Earth

As our planet coalesced in the early days of the Solar System, it was certainly not an idyllic parkland. A picnic would have been no fun on the infant Earth. This was no place to safely raise your child, grow plants, or walk your dog. Continent-sized chunks of ice and rock, such as the comet shown in Fig. 3.1, careened through the inner Solar System. Many collided with Earth. It must have been quite a show.

Each of these collisions was a double-edged sword. Huge craters were gouged in the molten crust of the young world; tectonic forces responded by exploding volcanic geysers into Earth's new sky. Any life that had taken hold on the new planet would be quickly extinguished by these catastrophic events. But at the same time, each collision brought new material from space. From comet impacts, Earth's primeval carbon-

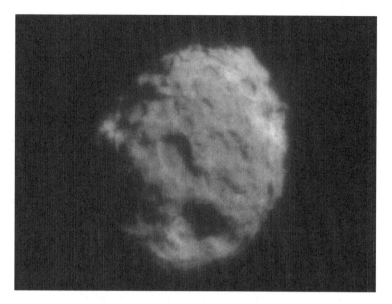

Fig. 3.1 The nucleus of Comet Wild 2, as photographed by NASA's Stardust mission (courtesy of NASA)

dioxide-rich atmosphere began to form. Comet-supplied water fell from the skies in huge quantities, only to be evaporated by the high-temperature crust.

Over hundreds of millions of years, Earth gradually cooled. But about 4.2 billion years ago, our world experienced the largest impact of them all. A world about the size of Mars, with about 10 % the mass of Earth, is thought to have broadsided our world.

Once again, Earth was in a molten state. Much of the debris was flung into space, where it ultimately coalesced under self-gravitation to become our Moon.

The population of comets and asteroids in the inner Solar System began to thin. Finally, Earth's crust was, at least in selected regions, solid rather than molten. Finally, all of the water that fell from the skies did not immediately evaporate.

About 3.8 billion years ago, life began. We do not yet know the entire story of this remarkable development. But life had arrived on the young planet. Earth, while not yet a park, was no longer barren.

Robots and Humans Enter Deep Space

So far, humans have walked only upon our planet's single natural satellite, the Moon. Our robots, however, have ventured farther. They have flown by every major world in our Solar System, orbited Mercury, Venus, Mars, Jupiter, and Saturn. One has

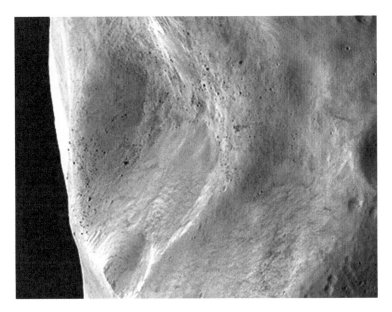

Fig. 3.2 Close up photo of asteroid Lutetia, snapped by Rosetta (courtesy of ESA)

penetrated a bit into Jupiter's alien atmosphere before the high pressure squashed it like an egg. Others have landed upon Venus, Mars, and Saturn's satellite Titan. A few have actually touched near asteroids and flown near the nuclei of comets. As we write, one is speeding towards a rendezvous with Pluto and another is serving as a roving robot geologist on the surface of Mars. Four emissaries from Earth—*Pioneer 10* and *11* and *Voyager 1* and *2* are departing the Solar System, moving into the vast emptiness between the stars.

We live in a golden age of Solar System exploration. Perhaps we take the explosion of knowledge regarding our celestial neighborhood for granted because much of the exploration is done by robots instead of humans. Robots are cheaper—they do not require elaborate life support, and they don't get bored during long-duration space voyages. But something is missing—the excitement of human explorers as they take the first steps on a new world.

During the early years of humanity's space program, all interplanetary ventures were conducted by only two nations—the Cold War adversaries the United States and Russia. But that is changing.

A host of international missions have now been conducted, are underway or planned that will greatly add to our knowledge of the Solar System. Using solar-electric propulsion (a solar-powered low-thrust rocket with higher exhaust velocity than chemical rockets), the U. S. Dawn probe is exploring several of the largest asteroids—Ceres and Vesta.

Not to be outdone, the European Space Agency (ESA) is conducting the Rosetta mission. In 2014, if all goes well, this probe is scheduled to deploy a landing craft to explore a comet's nucleus. It has already obtained close-up photographs of several asteroids (Fig. 3.2).

A few years ago, the Japanese *Hayabusa* (also called *Muses-C*) touched down on the asteroid Itokawa, retrieved some samples, and in a cliffhanger of a return mission, ultimately landed in Australia. Analysis of the few grams of the asteroid sample returned is currently underway.

Before too many decades have elapsed, it is almost certain that humans will follow in the footsteps of these robotic asteroid explorers. NASA and other space agencies are developing rockets and spacecraft that can support human expeditions to near asteroids and extinct comets. It will be necessary for crews to endure many months in weightless conditions in the high-radiation environment beyond the protection of Earth's magnetic field. But many astronauts and cosmonauts are sure to volunteer for the honor of exploring these small, celestial neighbors of Earth.

Beyond the Solar System

But how special is our Earth and Solar System? Since the 1990s, astronomers have refined their techniques to detect with confidence planets circling other stars in the Milky Way Galaxy. On Oct. 28, 2012, author Greg Matloff consulted the online extrasolar planet encyclopedia (http://exoplanet.eu) to check on the progress of planetary searches. On October 25, 2012, we knew of 843 extrasolar planets. There were on this date 665 confirmed planetary systems and 126 known multiple-planet systems.

Even though recent research points to planets attending Alpha Centauri B, a nearby Sun-like star, very few of the confirmed planets reside in the habitable zones of their stars—the regions in which our type of life could thrive on an Earth-like world.

Our abilities to detect distant Earths are improving rapidly, and we should before long be able to detect analogs to our home planet. But even the closest extrasolar planets are trillions of miles distant. Interstellar travel times using any known technology will be measured in centuries or millennia. If humans foolishly degrade their home world, there are no near, easily accessible Earth-like worlds on which we could start anew.

Further Reading

Solar System data are available from a wide variety of sources. One of our favorites is K. Lodders and B. Fegley, Jr., *The Planetary Scientists' Companion*, Oxford University Press, NY (1998).
Is Pluto a major planet or not? The popular astronomy press has done a good job of covering this debate. See, for instance, F. Reddy, "Top 10 Astronomy Stories of 2006," *Astronomy*, 2006; 35 (1) : 34-43.

The authors of this book survey prospects to ultimately travel to some of these distant worlds in *Living Off the Land in Space: Green Roads to the Cosmos* (Springer-Copernicus, NY, 2007). For a more mathematical review of this topic, consider Gregory Matloff's *Deep-Space Probes*, 2nd ed., Springer-Praxis, Chichester, UK (2005).

Chapter 4
Earth Before People: Utopia or Nightmare?

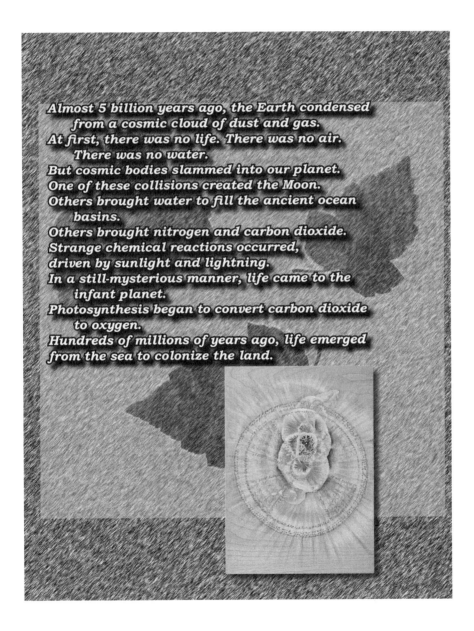

Almost 5 billion years ago, the Earth condensed
   from a cosmic cloud of dust and gas.
At first, there was no life. There was no air.
   There was no water.
But cosmic bodies slammed into our planet.
One of these collisions created the Moon.
Others brought water to fill the ancient ocean
   basins.
Others brought nitrogen and carbon dioxide.
Strange chemical reactions occurred,
driven by sunlight and lightning.
In a still-mysterious manner, life came to the
   infant planet.
Photosynthesis began to convert carbon dioxide
   to oxygen.
Hundreds of millions of years ago, life emerged
from the sea to colonize the land.

G. Matloff et al., *Harvesting Space for a Greener Earth*,
DOI 10.1007/978-1-4614-9426-3_4, © Springer Science+Business Media New York 2014

"O World! O Life! O time!
On whose last steps I climb,
Trembling at that where I had stood before;
When will return the glory of your prime?
No more – Oh never more!"
—From the poem "A Lament" by Percy Bysshe Shelley

In this chapter, we review the evolution of life on Earth and attempt to address the issue of how pleasant our planet's pre-human environment was. But before we begin, it might be helpful to investigate how benign or special early Earth was for the origin and evolution of life. We have learned a lot in the past few decades indicating that just having a small, wet planet in the Sun's habitable zone did not make our ultimate appearance a done deal.

Starting in the 1990s, astronomical technology had developed to the point at which small stellar motions caused by attending planets could be accurately monitored. It has become apparent that all or most Sun-like stars are accompanied by planets. But do not cheer too loudly. Many of these planetary systems are alien indeed and many would be less than beneficial for new or evolved life forms.

On October 29, 2012, one of the extrasolar-planet online data bases (http://exoplanets.org/table) showed that of the 631 confirmed worlds in this tabulation, 126 are classified as "hot Jupiters." Such a world has an approximate mass between half of Jupiter's (about 150 Earth masses) and 22 times the mass of Jupiter (about 6,000 Earth masses). The orbital period (year) of a hot Jupiter in this table is 10 days or less.

Not all Earthlike worlds are as fortunate as our home planet. If a hot Jupiter (Fig. 4.1) were in our Solar System, asteroids and comets would be deflected by the monster world from their stable tracks. Impacts on Earth would be far more frequent than they are. It would be hard for life to gain a foothold. Another problem would be constant changes in climate caused by the perturbation to our planet's orbit caused by the nearby monster world.

However, Earth was lucky. The inner Solar System is a relatively calm and stable place. Life formed and evolved over a period of billions of years. For eons, many species shared our planet's biosphere.

There can be no question that today, one species, *Homo sapiens*—bestrides the world. In some circles, it is fashionable to lament this situation. Has a golden age been lost and Eden been transformed into an omnipresent global civilization of commerce, popular culture, consumerism, and accumulation? Perhaps (according to some) Earth's prime is past, and the proper role of humanity is to hasten the degradation of this planet's natural environment.

It is easy to see from where such pessimism comes and how such defeatism has evolved. Our environment is degrading, impacted by human-caused pressures of over-population, rapid industrialization, and habitat destruction. However, one has to ask if the peaceful Eden of the theologian and the utopia of the philosopher ever actually existed.

Fig. 4.1 Artist's conception of a "hot Jupiter." Don't bet on life in this world's planetary system! (courtesy of NASA)

To investigate these questions, let's pick up the story after the appearance of the first life on Earth. In this chapter, we will attempt to describe billions of years of history as a narrative, often claiming as fact scientific hypothesis in many areas of evolutionary biology, planetology, and history. Although it may be that some of these events did not occur exactly as we describe, they are nonetheless reflective of many prevailing scientific theories.

Firstly, Earth had an atmosphere in this era—but oxygen was a rarity. The comet- and volcano-supplied early terrestrial atmosphere may have been rich in nitrogen and carbon dioxide, and there may have been ample hydrocarbons, as in the modern atmosphere of Saturn's satellite Titan (Fig. 4.2), but free oxygen was either very rare or non-existent.

Early terrestrial life not only survived without oxygen; these anaerobic (oxygen-hating) forms actually thrived in their environment. But life operates according to Darwinian evolution. Environments change, organisms can mutate, and the descendants of those best suited to an altered environment come out on top.

Around 3.8 billion years ago it seems a crucial mutation occurred. Perhaps challenged by more successful forms, a microscopic anaerobe that had retreated to the upper layer of Earth's young ocean learned to gather energy from sunlight. In this process of photosynthesis, this plant-like organism converted solar energy and carbon dioxide into glucose, producing oxygen as a waste product. Over time, this new life form began to thrive.

Fig. 4.2 A mosaic of Titan's surface, as photographed by the descending European Space Agency (ESA)/NASA Huygens probe (courtesy of NASA)

This was perhaps the most critical of Earth's natural environmental catastrophes. The anaerobic forms must have felt that their Eden was slowly becoming a sewer or a toxic waste dump as oxygen became more and more prevalent in the atmosphere. Perhaps with a slight sense of desperation, they retreated to volcanic vents on the deep ocean floor, only to trouble the surface world during periodic, noisome El Nino events.

The surface layers of Earth's oceans now belonged to oxygen-loving microscopic, one-celled life forms. Competition and predation must have been fierce. As Earth slowly developed its oxygen atmosphere, mutation and natural selection gave rise to multi-cellular colony organisms similar to modern jellyfish.

It was thought for many years that the crucial step from single-cell to multi-cell life was very time consuming. After all, terrestrial life began about 3.8 billion years ago. The earliest fossil evidence of multi-cell life dates to about 500 million years ago.

However, a triumph of "small science" announced in early 2012 indicates that the evolution of multi-celled life was a lot easier than had been suspected. The remarkable results of this very straightforward experiment are summarized in Press Release 12-009 of the U. S. National Science Foundation (January 16, 2012). This document can be accessed on-line at www.nsf.gov/news_summ.jsp?cntm_id=122828.

A research team consisting of Will Ratcliff, Michael Travisano, Ford Dennison, and Mark Borrello at the University of Minnesota started with a sample of common Brewer's yeast (saccharomyces). They added the yeast cells to nutrient and allowed them to grow for 1 day. Then, they spun the yeast in a centrifuge to stratify the material by weight. Clusters of cells tended to fall to the bottom.

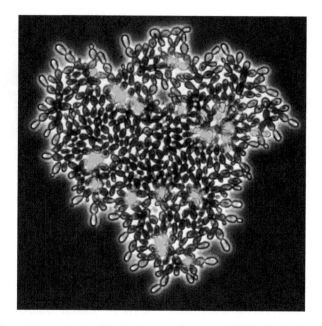

Fig. 4.3 Micro-photograph of multi-cellular yeast (courtesy of NSF. Credit: Will Ratcliff and Mike Travisano)

The samples were repeatedly agitated in this manner for 60 days. The results were snowflake-like structures of hundreds of cells. After cell division, related cells tended to remain attached (Fig. 4.3). According to the NDF press release, multi-celled colony organisms may have evolved independently as many as 25 times during the history of Earth's biosphere.

Some of these colony organisms, perhaps to avoid being devoured by their stronger, nastier neighbors, developed specialized cells that acted as hard, protective shells. This was the great era of the trilobites and other early shellfish. One of the few present-day survivors of this period is the horseshoe crab.

Hardy plants were beginning to spread from the ocean shore inland. And insects that played the role of pollinators accompanied them. Vast forests covered the planet. The buried organic remains of this so-called "Devonian Era" formed over tens of millions of years and were converted by geological processes into coal and oil. We are consuming this wealth of our planet at a prodigious rate. It will be gone within a very few centuries. And the geologically stored carbon released in a mere instant (from our planet's point of view) currently threatens global climate stability.

However, evolution in the oceans continued at an ever more rapid pace. Somewhere at sea, a shellfish mutated to develop a crude interior skeleton. The first armored fish—the ancestors of modern sharks—had evolved. Nature's new trick, the backbone, soon caught on. The seas filled with vertebrates.

By about 300 million years ago, Earth's plant-produced oxygen layer was fully operational. In the stratosphere, solar radiation energized chemical reactions that converted some oxygen molecules into ozone. With an ozone layer in place to filter harmful ultraviolet sunlight, the stage was finally set for vertebrates to emerge from the sea.

Everywhere, the strong preyed upon the weak. Civilized protections were absent in the natural state. In tidal pools along the shores of Earth's oceans, small fish took refuge from their fierce neighbors.

As the Moon orbited Earth and Earth spun on its axis, lunar and solar tides periodically varied water levels in these fragile habitats. Somehow, a fish developed crude lungs so it could spend a portion of its existence above sea level—and therefore survive extreme tidal variations in water level. This very beneficial mutation was passed to this creature's offspring. Amphibians began to emerge from the seas. In the fullness of time, some of these creatures were altered so they could live full time on land. These were the ancestors of reptiles, dinosaurs, feathered (birdlike) dinosaurs, and mammals.

Around 200 million years ago, the warm forests of Earth belonged to the big guys. Huge plant-eating dinosaurs browsed on the vegetation. Smaller, smarter carnivores—the raptors that would give rise to modern birds—silently stalked the huge ones.

Hiding in the undergrowth, doing their best to avoid their ferocious, gigantic neighbors were small, rat-like mammals. These—the distant ancestors of elephants, tigers, and humans—emerged briefly from their warrens to grab a dinosaur egg or two, and then scurried for cover.

Life was abundant in these times, and life was diverse. But at least from the viewpoint of the mammals, life was no picnic, and Earth was no Eden. And so it might have remained, as Earth spun upon its axis and danced around its star. But change seems to be built into all natural systems. After more than 100 million years of stability, catastrophe rained from the skies.

Even without human intervention, bad things happen to our planet's biosphere. At intervals of tens of millions of years, occasional mass extinctions occur. During these events most terrestrial organisms perish, and many terrestrial species disappear. When the dust finally settles, a remade biosphere emerges, one with new life forms and new ecological niches.

Some of these events are perhaps produced by the eruption of super volcanoes. Compared with these, historical eruptions such as Krakatoa, Vesuvius, and Thera seem like mere sneezes. Volcanic dust from these blow-ups, suspended for decades in Earth's upper atmosphere, would block sunlight and result in unending winters.

Other mass extinctions in pre-history seem to have cosmic causes. Certain types of nearby exploding stars might have bathed our planet in a sea of gamma rays, fatally irradiating all but the hardiest specimens of terrestrial life.

About 65 million years ago, as the herbivores grazed under the watchful eyes of the giant raptors, as tiny mammals briefly emerged and immediately retreated for cover, another type of celestial catastrophe occurred. Perhaps some of the doomed

Fig. 4.4 The great impact at the end of the Cretaceous Era would end the reign of the dinosaurs but open the way for mammals, including humans. Perhaps the event registered on some dinosaur brains before they were snuffed out (courtesy of NASA)

animals glanced at the brightening sky, unaware of what the celestial spectacle signified.

From the sky, a chunk of ice or rock descended towards Earth's surface. This 10-km wide fragment of asteroid or comet streaked across the sky and toward the ground at more than 10 km/s. In the dim brains of the doomed dinosaurs, the aerial show must have seemed like a second Sun, dashing rapidly across the sky (Fig. 4.4). The explosion was like a million hydrogen bombs igniting in the same place and at the same time. A towering mushroom cloud that reached to the stratosphere replaced the enormous fireball. Shockwaves raced across the plains and through the forests. Enormous firestorms devoured most vegetation. Towering tsunamis raced through the oceans, radiating outward from ground zero.

Various seismic events—volcanoes and earthquakes—compounded the damage to living organisms, even those separated by geography from the direct effects of the blast. Higher life, at least, was erased from most of what would later be known as Earth's western hemisphere.

Earth was enshrouded in a vast halo of dust. For years or decades, temperatures plummeted globally as the dust layer reflected sunlight back into space. Most surviving vegetation perished with the onslaught of the cold, followed by most of the herbivores who fed upon it. Carnivores who had survived the impact may have initially had a field day, feeding upon the bodies of the deceased plant eaters. But they died as well, since their food source could not be replenished.

Decades after the impact, skies began to clear to reveal a greatly altered world. Perhaps because they could hibernate, perhaps because small size had allowed them to find shelter, or perhaps just because they were lucky, a handful of mammals emerged. They were accompanied by the ancestors of modern birds—small, feathered flying dinosaurs—who survived perhaps due to the thermal insulation of their feathers and their ability to fly from a degraded environment to a better one.

Our distant ancestors may have dominated the landscape in the early Tertiary period, as buried seeds began to sprout to re-germinate Earth's ruined forests. But life was hard—it was no picnic, and again, it was not Eden.

As the eons flew by, the mammals mutated to fill ecological niches vacated by their vanished rivals. Creatures much like elephants, whales, deer, and apes ultimately evolved. As they competed for food and territory, evolutionary pressure picked up.

A few million years ago, an arboreal ape-like creature had evolved in central Africa. With opposable thumbs and a comparatively large brain, these creatures were well suited to life in the canopies of tall forest trees.

However, the environment changed again. As forests were replaced by prairie, some of these organisms descended from their perches to attempt life on the hostile environment of the surface. They faced great odds as they attempted to avoid the predations of the great cats and others. They could not know how momentous their descent from the trees was, but these humble beings were the ancestors of us all. In the rough-and-tumble environment of the African savannahs, competition would spur tool use and a rapid increase in brain size.

A few hundred thousand years ago, the first true humans evolved. No longer would they be victims of environmental change; the fate of the entire terrestrial environment would ultimately rest in their hands instead.

Further Reading

For a very readable survey of life's origin and development on our planet, see John Reader, *The Rise of Life*, Knopf, NY (1986). The devastating impact that ended the reign of the dinosaurs is discussed by Carl Sagan in *Pale Blue Dot*, Random House, NY (1994).

Part II
Paradise Lost?

Chapter 5
The Environmental Dilemma: Progress or Collapse?

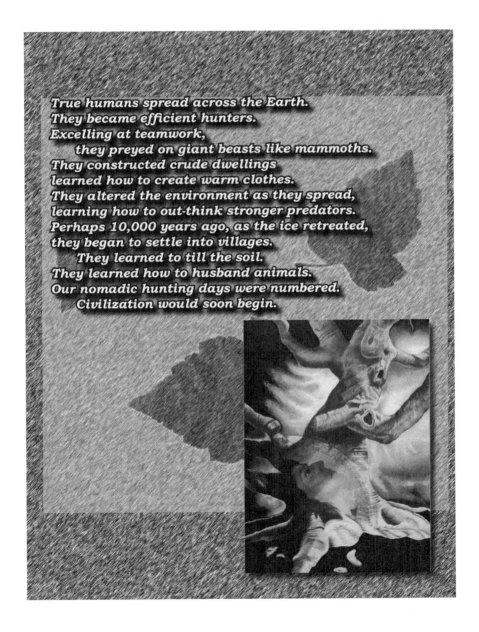

True humans spread across the Earth.
They became efficient hunters.
Excelling at teamwork,
    they preyed on giant beasts like mammoths.
They constructed crude dwellings
learned how to create warm clothes.
They altered the environment as they spread,
learning how to out-think stronger predators.
Perhaps 10,000 years ago, as the ice retreated,
they began to settle into villages.
    They learned to till the soil.
They learned how to husband animals.
Our nomadic hunting days were numbered.
    Civilization would soon begin.

G. Matloff et al., *Harvesting Space for a Greener Earth*,
DOI 10.1007/978-1-4614-9426-3_5, © Springer Science+Business Media New York 2014

"O me! O life! Of the questions of these recurring,
Of the endless trains of the faithless, of cities filled with the
foolish,
Of myself forever reproaching myself, (for who more foolish
than I,
and who more faithless?)"
—From the poem "O Me! O Life!" by Walt Whitman

As we have seen, the distant past was difficult for our ancestors. But where have the development of civilization, the agricultural revolution, the art of metallurgy and the scientific revolution left us? We seem to be suspended from an environmental cliff. We can't hang on, but we are doomed if we let go!

Our global civilization may be the first to understand the dilemma of the environment. But we are not the first to be threatened by environmental collapse. Rather than endlessly reproaching ourselves about the foolishness that led to our predicament, it may be helpful to briefly review humanity's prehistoric and historic interaction with the environment.

Early in human prehistory, our environmental footprint was minimal. For about 2 million years, our ancestors lived off whatever nutritious local vegetation was in bloom, ate meat from whatever they could catch or steal from larger predators, and did their best to avoid the jaws of the bigger, fiercer animals. Without permanent habitation, it was not hard to escape a degraded environment. Effectively nomads, they would simply pack up shop and move. But perhaps because we were slower and less well armed than large cats, bears, or wolves, human progenitors began to depend more on cleverness and stealth than on physical strength. As our brains developed, so did our tools.

With the aid of spears and arrows, we developed the skills to cooperatively hunt large game and to kill at a distance. The unknown genius that first tamed fire opened the way to the ultimate spread of humans from the tropical African home to temperate and arctic climes.

Perhaps their tribe was pursued by more powerful neighbors; perhaps it was population pressure or environmental change. Whatever the cause, a small group of humans found itself hemmed in by the ocean, on what is now the southeastern coast of Asia. More than 40,000 years ago one of them might have observed that fallen trees floated in water after a storm. Following this person's keen observation, the tribe strung logs together to construct the first ocean-going rafts. Island hopping across the shallow sea, the people of this tribe ultimately arrived in what is now Australia.

Strange new animals inhabited a landscape covered with unfamiliar vegetation. After a few generations, these progenitors of the modern aborigines learned by observation how effective fire was in improving fertility and clearing undergrowth. It may have been an early tool to improve the hunting ground by thinning the forest. For the first time, humans were setting controlled fires to alter their environment. Ultimately, this would lead to what became "slash and burn agriculture." It represents what may have been the first large-scale manipulation of the environment by humans.

In other parts of the world, as climate cooled and glaciers advanced, human ancestors became well adapted to the Ice Age. From cave paintings in northern Europe, we have a fair idea about methods used by hunting bands to bring down large game animals such as the now-extinct mammoth.

About 13,000 years ago, human bands migrated from Asia across a narrow land bridge connecting Siberia and Alaska. Modern humans had invaded the Western Hemisphere and would soon claim it for themselves. Interestingly, biodiversity in the New World declined a bit as certain large-animal species became extinct, possibly due to the hunting skills of the new settlers.

Then, the climate began to warm. The Bering-Sea land bridge was submerged as melting glaciers raised ocean levels. In Europe, Ice Age people were faced with a major dilemma. As the ice retreated, so did the big-game herds that the paleo-Europeans depended upon. Unless people were to follow the herds, radical alterations in life style were required.

Not all near-humans were able to adapt to the warming climes. Our close cousins, the Neanderthals, vanished with the retreating glaciers. It may have been with timidity, fear or trepidation, that the once dominant (male) hunters approached the more submissive (female) gatherers about possible solutions. But about 10,000 years ago, in the Middle East and central Turkey, something brand new began to spread across the face of Earth.

The Agricultural Revolution

This was the agricultural revolution, which began at the dawn of the Neolithic period—the New Stone Age. For the first time, in Jericho, Catal Huyuk, and a host of other sites, humans were putting down permanent roots. We learned how to farm, and how to husband domestic animals. No longer were we at the mercy of the seasons, always pursuing the animal herds.

However, there were other problems as population of the early towns rose to the hundreds and thousands. One problem were pre-Neolithic nomadic bands, who might be tempted to rob our ancestors of Earth's bounty. So defensive walls were constructed around the growing towns. Organized warfare, which may have existed for millennia, became a major human endeavor.

There were other problems as well. In the beginning at least, early agriculturists were at the mercy of the elements. Too little rain would parch the crops; too much rain would drown them. So a mythology developed that connected humans to nature.

There were male sky gods such as Zeus, Jupiter, and Thor. They were responsible for the defense of the growing settled communities against outside intruders. Other male gods—underground deities such as Poseidon and Neptune—were responsible for large-scale catastrophes such as earthquakes, volcanic eruptions and tsunamis. Elaborate rituals were devised to keep these dangerous beings at bay.

This was also the era of the Great Goddess. Female deities were in charge of the monthly and seasonal cycles. The greatest of these was Gaia, the goddess of Earth. Until about 1,000 B.C., when the Age of Stone had yielded to the Bronze Age and was itself surpassed by the Iron Age, most civilized humans felt tied to the cycles of nature. Some of the great stone-circle observatories they used to keep track of seasonal cycles—notably Stonehenge in the United Kingdom—can still be visited today.

In this long era, people began to utilize the powers of nature in an unprecedented fashion. The early log rafts of the Paleolithic Australian migrants were rendered obsolete when some budding navigator observed the behavior of large water birds, perhaps swans, on a river such as the Nile in Egypt. If such a creature desires to move against the river's current with the wind at its back, it simply fluffs its feathers and glides across the waters.

By 4,000 B.C., crude sailboats carried trade goods between communities located along the Nile. Soon, some of these ventured into the Mediterranean. The Cycladic and Minoan civilizations further developed the sailing craft—with innovations including the keel, which allows sailboats to tack against the wind. Before the end of the Bronze Age, some of these craft ventured out into the Atlantic Ocean, to visit areas that are now England, Scotland, Wales, and Ireland. Before the Golden Age of Athens and the rise of Rome, crews of other vessels were exploring and trading along the coast of Africa. Sometime around 2,000 B.C., the death-knell of this era resounded when someone on the Asian Steppes noticed that the horse, a delicious game animal, could be tamed for other functions.

The Age of Iron

The new partnership of human and horse opened the way for the ascendancy of the male, warlike sky gods. Carrying the banners of these deities and sophisticated weapons of bronze (and later of iron), chariot-equipped armies fanned out from their central- Asian homelands to dominate the Indo-Pakistan subcontinent and what is now Greece. Myths about the hybrid human-horse centaurs reflect this turbulent era of prehistory.

City-states began in certain parts of the world, to be incorporated into a new form of human organization—the centralized empire. Beginning in Africa and Asia with the rise of Egypt, China, and Babylon, the cult of the conquering warrior ultimately spread throughout Europe.

At least some of this centralization was a response to human civilization's interaction with the environment. Along the Nile River, human settlements grew in a narrow band of fertile land that was nurtured by the river and always threatened by the encroaching desert. As agriculture succeeded, human lifespan increased and the population expanded. Cities required rivers as both a fresh-water source (upstream) and a sewage dump (downstream).

This arrangement worked fine as long as there were few cities. But as human habitation expanded along the Nile, a centralized government was necessary to regulate water use for irrigation and waste disposal. The great Egyptian empire, which was instituted more than 5,000 years ago and served as a model for many future states, was very likely a response to environmental pressure.

As the ancient empires flourished, warred with each other, and were ultimately incorporated in the all-encompassing empire of Rome, people may well have noticed a change in the human relationship with the world around them. No longer were people totally at the mercy of the elements. Humans were supreme, dominating the landscape as well as other life forms.

Sadly, Iron Age humans did not always use their new powers for humane purposes. An early environmental outrage occurred when Rome, after having finally defeated Carthage in the Punic Wars, sowed the soil around Carthage with salt in 146 B.C. to destroy the land's fertility.

There were setbacks to human ascendancy of Earth of course—earthquakes, volcanoes, and possible meteorite impacts—but the devastation was local. With the rise of Christianity and Islam, the powers of humanity grew, as did the population.

One problem faced by the expanding cities was the diminishing local supplies of firewood. People required this resource for heat in winter and to cook food. But forests mature over periods of decades or centuries, so growing populations needed to travel farther and farther from home to obtain firewood.

During the Middle Ages, someone in England, Ireland, or Scotland realized that peat, an organic-rich soil, could be dried and used for fuel. Before A.D. 1600, though, peat had been supplanted in cities such as London by a new fuel—coal.

Early Air Pollution Episodes

The new fuel had the advantage of freeing the growing population of London from a dependence on diminishing local forests. But this societal advance was a two-edged sword. Coal was generally burned in household hearths. Scrubbers and other pollutant-removal techniques were unheard of at the time. Soon, an omnipresent dark cloud reduced sunlight levels. Exposed structures open to the elements began to feel the effects of corrosion. Worst of all, humans began to suffer from a host of respiratory ailments that had previously been rare or unknown.

As we now understand, atmospheric processes render air pollution a non-local phenomenon. Fallout from the sulfur-bearing effluent clouds produced in the cities traveled into the countryside, affecting humans, crops, and other organisms far downwind from the concentrations of coal-burning hearths and furnaces.

This type of air pollution—characterized by sulfur oxides and particulate matter is fittingly called "London Smog," after the city in which it first became evident. Modern centralized coal-burning power plants overcome much of this problem by employing scrubbing and filtering techniques to remove most of the sulfur

Fig. 5.1 Representation of a hear engine, showing the extraction of useful energy as heat flows from a higher temperature to a lower temperature

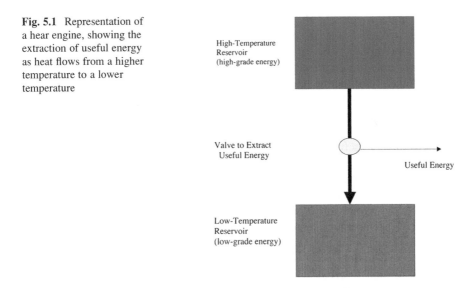

High-Temperature
Reservoir
(high-grade energy)

Valve to Extract
Useful Energy

Useful Energy

Low-Temperature
Reservoir
(low-grade energy)

compounds and particles from the effluent plume. But there are limits to the effectiveness of pollution control, no matter how sophisticated the technologies we employ.

The Modern World: Running Up Against Environmental Barriers

Four centuries after the large-scale introduction of fossil fuels (coal and later, petroleum), our global civilization runs on these hydrocarbons. These remnants of ancient forests, modified by geological processes for millions of years, are being consumed by our global civilization at an ever-increasing rate and will be completely gone within a century or two.

In part because of this comparatively plentiful and inexpensive energy source, a larger fraction of humans live better than ever before. Our life spans are longer, and a greater percentage of children (at least in the developed world) survive to adulthood than in any previous era.

However, the laws of nature conspire to limit the longevity of our consumptive global civilization. To appreciate what's going on, we must make a short detour into that branch of classical physics called thermodynamics. Thermodynamics basically means "motion from heat." It essentially describes how to obtain useful work from an idealized heat engine.

Figure 5.1 is a schematic representation of such a device. Energy flows downhill from a hot reservoir to a cold reservoir. Think, for example, of your home furnace or an automobile propelled by an internal combustion engine. Fuel is burned at as high a temperature as possible (the "hot reservoir") and waste products of this combustion are expelled to the much colder environment (the "cold reservoir"). Somewhere in this process, some of the energy produced by the combustion is diverted to heat your home or turn the auto's wheels.

The first law of thermodynamics is often stated in gambler's parlance as "you can't get something for nothing." This means that any heat engine functions in the real world—all useful energy supplied by such an engine comes from heat. This is reasonable; otherwise, a heat engine would be a perpetual motion machine, magically supplying useful energy without fuel.

However, the second law of thermodynamics is even more restrictive: it states, again in the language of the gambler, "you can't even break even." No heat engine will ever be 100 % efficient. Some waste heat will always be expelled to the environment. From a physics point of view, efficiency increases as the hot-reservoir temperature increases and the cold-reservoir temperature decreases.

Many contemporary fossil-fuel power plants have efficiencies of around 50 %. Since nuclear-fission plants often expel waste heat into enclosed water-filled "cooling towers" to avoid radiation release to the environment, they are generally slightly less efficient than fossil-fuel power plants that expel their waste heat to cooler natural water bodies.

In order to reach efficiencies as high as 70 %, it is necessary to employ the technologies of magnetohydrodynamics (MHD). In experimental MHD plants, the high-temperature reservoir operates at about 1,000 Kelvin (K). At such high temperatures, the circulating fluid is an ionized gas or plasma rather than superheated steam. A major obstacle to the wide-scale utilization of high-efficiency MHD technology is the very real possibility of injury or death if an accident caused the high-temperature plasma to escape into the environment.

The second law of thermodynamics is often called "entropy," which means that the disorder of the universe is always increasing. For the entire universe, this can be visualized by realizing that all potential for (high-grade) energy production was concentrated in the interior fusion-fuel sources for infant stars in the universe's early ages. As the universe winds down in many billions of years, most of this energy will have been transferred to (low-grade) random motions of interstellar molecules and atoms—heat.

When applying the second law of thermodynamics to life, one realizes that considered in isolation from its environment, a life form is negentropic; that is, it learns more and remembers more as it matures. But this increasing order comes at the expense of the environment. Fortunately, Earth is not a closed system, and it is bathed with enormous amounts of energy from the Sun every day. It if from this sunlight that life derives the energy necessary to create and maintain itself.

Also, localized resource depletion is a recurring theme through most of human civilization. If you consult the Bible or Homer's epics, you will learn that in the Bronze Age, Lebanon and Crete were famous for their old-growth forests. But 4,000 years of high civilization has seriously degraded the natural environments of these and other ancient sites. Closer to home (in time if not in space) the paleo-American cliff dwellings in the American Southwest were abandoned about

Fig. 5.2 The full Earth, photographed by an Apollo crew between Earth and the Moon (courtesy of NASA)

1,000 years ago. Increasing population put too much of a strain on the limited water reserves in this arid environment for that culture to thrive and prosper.

In ancient times, Earth's population was considerably lower than it is today. It was possible to think of the planet as an infinite sink for pollution. If your environment became degraded, it was readily possible to find another home elsewhere. But in today's heavily populated global society, we do not have the luxury of picking up roots, abandoning our cities, and migrating to fresher climes.

Although our current environmental crisis was generations in the making, it came into clear focus in the 1970s. This may well be due to the Apollo photographs of Earth from deep space, such as the one reproduced as Fig. 5.2. For the first time, Earth was recognized as a fragile oasis of life suspended in a mostly sterile cosmos, a precious jewel rather than an infinite sink for civilization's waste products.

Today, much environmental debate is centered upon the very real problem of global climate change. But there may be other global effects of great significance. One is ocean pollution. We had perhaps mistakenly thought that this is a non-issue, since most industrialized regions have efficient sewage disposal systems and regulate industrial river and ocean pollution.

About 70 % of Earth's surface is ocean. Most of the planet's oxygen, which is required for the survival of humans and other animals, is produced by microscopic organisms in the warm upper layers of the oceans. But recent research indicates that our global, technological civilization may be threatening the survival of oceanic life.

As shown in Fig. 5.3, all oceans feature large, circular ocean currents called gyres. The most famous gyre is the Gulf Stream, in the northern North Atlantic. Oceanographic expeditions to regions within the North Pacific gyre have revealed a

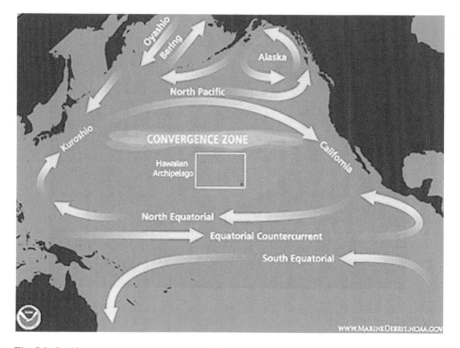

Fig. 5.3 Pacific ocean currents (courtesy of NOAA)

disturbing development. Apparently, plastics do not rapidly degrade in the ocean environment. When storms flood coastal cities, plastic debris often runs out to sea. This is broken down into small particles by wave and wind action. These accumulate in regions within the gyres, such as the "Pacific Garbage Patch."

This would not be of great significance were it not for the oceanic food chain. Small organisms mistake the polymer particles as food and ingest them. Other species eat the small creatures up to and including many significant fish for human consumption. It is unclear at present exactly how large the oceanic garbage patches are. They are not easily observable from space. In situ monitoring may be necessary. Further, ocean cleanup techniques are cumbersome (Fig. 5.4).

However, a rational civilization concerned about its future can approach problems like this as they are recognized. One possibility, already underway on a small scale and sure to accelerate, is to replace stable polymers in plastics with organic, plant-grown substances that readily degrade in the environment.

We are no longer completely at the mercy of the environment. The truth is that humans, as the stewards of Earth, must develop an enlightened approach to the consideration of environmental issues if our civilization is to continue to thrive.

One approach is to broaden our definition of environment to include the solar vicinity of our planet. If energy and other resources can be obtained from space and some of the waste products of civilization (heat, at least) disposed of in the sterile environment above the atmosphere, the chance exists to greatly enrich the living standards of most people, without jeopardizing Earth's biosphere. But accessible

Fig. 5.4 An ocean cleanup operation (courtesy of NOAA)

space is not infinite, as was demonstrated by the collision of two near-Earth artificial satellites in early 2009 and the resulting cloud of orbital debris.

Instead of viewing the immediate future with gloom, perhaps some optimism can be generated. If we use the tools at our disposal and rationally plan for the future, the human course may be bright indeed.

Further Reading

For a very readable account of human prehistory and history between 35,000 B.C. and A.D. 500, check out Jacquetta Hawkes, *The Atlas of Early Man* (New York: St. Martin's Press, 1976). Of the many books that treat historical and modern pollution episodes, one nice choice is Laurent Hodges, *Environmental Pollution*, 2nd ed. (New York: Holt, Rinehart and Winston,1977).

Many sources discuss application of thermodynamics to energy and environment. A fairly up-to-date and beautifully illustrated text considering this topic is G. T. Miller, Jr., *Environmental Science*, 4th ed. (Belmont, CA: Wadsworth Publishing, Belmont CA, 1993).

A fascinating book contributing to the study of why some civilizations fail while others endure is Jared Diamond, *Collapse: How Societies Choose to Fail or Succeed* (New York: Viking, 2005). Civilizations are compared in terms of international relations, environmental impacts, internal politics, and other factors determining stability.

Chapter 6
An Exploding Population

*In the deserts of Arizona space habitat technology
has been coming to Earth
Italian-born architect Paolo Soleri has been
    constructing an Arcopolis.
When completed, thousands of people could
    inhabit this site.
They will live in an energy self-sufficient
community equipped with wind turbines and solar
    cells.
Because work space will be close to residential
facilities they will bike or walk to work rather
    than drive.
Under the bright Arizona Sun, they will grow most
    of their food using hydroponics.
Recycling will be the norm-their carbon footprint
    will be minimal.
The Arcopolis may be the model for how terrestrial
billions will live in the future*

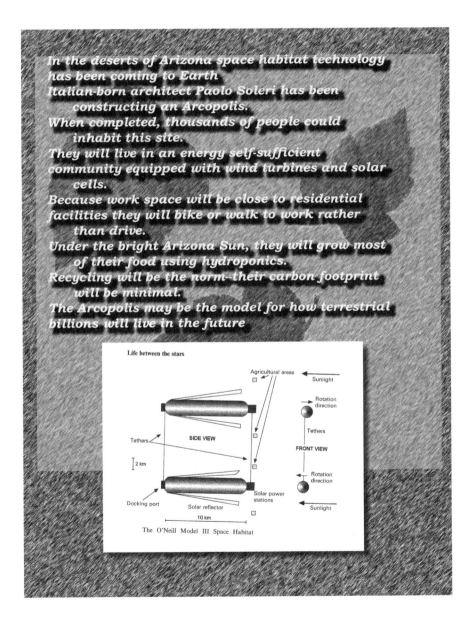

The O'Neill Model III Space Habitat

G. Matloff et al., *Harvesting Space for a Greener Earth*,
DOI 10.1007/978-1-4614-9426-3_6, © Springer Science+Business Media New York 2014

Habitats in Space, Habitats on Earth

"Crowds of men and women attired in your usual costumes,
how curious you are to me!
On the ferry-boats the hundreds and hundreds that cross,
returning home, are more curious to me than you suppose,
And you that shall cross from shore to shore years hence are
more
to me, and more in my meditations, than you might suppose."
—From the poem "Crossing Brooklyn Ferry" by Walt
Whitman

As Walt Whitman watched the multitudes commuting between Brooklyn and Manhattan in the days before the construction of the Brooklyn Bridge, he dreamed of the multitude that would make that crossing in the future. He could not have imagined the much larger population that has been born in recent years to challenge the future of Earth.

It is not uncommon for science authors to treat the population explosion as a form of plague, and to consider so many billions of humans a threat. Here we take the opposite tack: this is a challenge to be overcome. But if necessary changes to human lifestyle are made, healthy and productive lives are possible for the burgeoning human population. The opportunity exists to make these changes without desecrating our planet.

One thing to remember is that population levels have varied throughout history and prehistory. The number of humans existing at any time in our species' history has a lot to do with our interaction with the environment.

Human Population: The Prehistorical/Historical Record

Thousands of years ago, before humans spread around the globe from their homes in Africa, no one would have said there were too many people. In fact, the opposite may have been true. Because of the comparatively low level in human genetic variation, the original human population may have been very low. Perhaps we are all descended from a few thousand individuals.

Many things could reduce the life span of an early human: death in childbirth due to poor sanitation, infant mortality due to infectious disease, culling of the young and elderly by hungry predators, and so on. The entire human population 100,000–200,000 years ago was centered in Africa, and we numbered fewer than 100,000 individuals.

When fire was tamed a bit later, people could migrate to cooler climes and increase their numbers somewhat. The staple diet now consisted of cooked meat, which may have slightly increased the human life span since most food parasites are killed by cooking. But there were other hazards to contend with; nontropical winters must have been fierce during periods of global glaciation.

By the late Paleolithic Era, the human population must have numbered somewhere in the several millions. Even though people lived in small hunting groups of one hundred or so and the landscape was sparsely populated, humans spread around the world.

A significant increase in human population occurred with the onset of agriculture, around 10,000 years ago. Some of the Neolithic villages must have numbered in the thousands. Farming and animal husbandry meant that people living in settled communities enjoyed a more stable food supply than did their nomadic cousins. This must have resulted in a slight increase in average human longevity, but very few people reached the age of 50.

Metallurgy came into use with the Bronze Age, which extended from about 3000 to 1000 B.C. At least some people now lived in true cities with populations of 10,000 or more. With such large populations and metal tools, sanitation became a major endeavor. Delivering fresh water to a population center and removing sewage became the responsibility of a new class of civil engineers. Organized medical procedures also date from this period, and this also contributed to increased longevity. Famines in this period were alleviated as the new god-kings learned to store grain in good years and distribute it to their subjects when crops failed. There were as yet no global population counts, but tens of millions may have inhabited our world.

With the arrival of the Iron Age at about 1000 B.C. and the rise of the empires, population levels in major cities may have exceeded 100,000. Even with advances in hydrology, sanitation, and food distribution, a new limiting factor arose to reduce human population levels. As yet, there was no understanding of microscopic pathogens. Various plagues afflicted the human population, which were more effective than warfare in limiting human numbers.

All through the Middle Ages, the human population slowly increased, except when incidents such as bubonic plague—the so-called Black Death—caused human numbers to crash. But the total global population never exceeded a few hundred million.

Medieval Europeans were partly responsible for the spread of the bubonic plague. In a frenzy of irrational brutality, the Church-led Inquisition had slaughtered multitudes of women as witches. Many cats were also thrown into the flames, suspected of being witches' familiars. If these blameless creatures had survived, they might have killed more of the rats that hosted the parasite responsible for the deadly disease.

However, the fundamentalist fanatics did not have the last word. Around 1500, rationality reentered the human arena with the Renaissance. As classical learning was rediscovered, the scientific method was perfected. For the first time, a philosophical formalism existed whereby hypotheses about nature would be tested by observation and experiment.

Although the broadening of humanity's worldview by the outward-peering telescope pointed the way to the new science of physics and its attendant, world-creating technologies, it can be argued that the biological revolution spurred by the microscope was even more profound. As early biologists used this new tool to observe the behavior of organisms too small to be seen by the unaided eye, a new theory of disease began to take hold. As this germ theory matured, it became impossible to maintain the ancient concept that disease was caused by improperly balanced humors within the afflicted organism. Microscopic entities—first bacteria and later viruses—were pinpointed as the causes of disease.

By the year 1900, the human life span had taken another leap forward. Sterile conditions in the operating room did wonders to extend the life of typical humans in developed countries. No longer would a young woman expect to die in childbirth; no longer would a family have six children so that two might survive to maturity. Around this time, the human population approached or exceeded the 1 billion mark.

The twentieth and twenty-first centuries have witnessed an explosive increase in human population. In early 2009 there were more than 6 billion people on Earth. The population may peak at around 10 billion before 2050. This is a remarkable development. Even in light of destructive wars (such as World War II, which resulted in about 60 million deaths), newly evolved plagues such as AIDS (which has killed tens of millions), and many wars including episodes of "ethnic cleansing," the population continues to increase.

Population Increase: Its Causes and Ramifications

It has been pointed out that human population levels have followed a J-shaped curve—nearly flat for thousands or millions of years, followed by a very rapid increase, with a new, higher equilibrium level achieved at the end of the rapid increase. An example of such a J-shaped curve is shown in Fig. 6.1. Will population remain constant at around 10 billion, continue to increase, or crash to a much lower value?

One reason for the dramatic rise in human population is the success of modern medical science. With our current understanding of the evolution and behavior of human pathogens, drugs have been developed that have dramatically increased the human life span. Unhealthy behaviors have been altered (at least in the developed world), sterile medical procedures are the global norm, and knowledge regarding nutrition and nutritional supplements has become widespread. Another success of civilization has been elimination or reduction of the larger predators who once preyed upon us.

Population dynamics have a number of interesting aspects. One might think, for instance, that an average married couple must have two offspring to maintain a constant population. However, a certain percentage of every generation will not marry, and some children will die before they are physically mature enough to reproduce, even in developed countries. For these and other reasons, married couples must have an average of about 2.3 offspring to maintain a constant population.

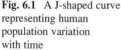

Fig. 6.1 A J-shaped curve
representing human
population variation
with time

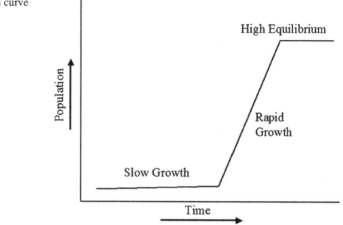

In the developed world, where there are many employment opportunities for women outside of the home, typical couples often have less children than the number required for a stable population level. In fact, recent demographic data shows that for much of the developed world (North America, Europe, and Japan), the birth rate has declined to the point where native populations are actually decreasing.

According to the United Nations, Europe is projected to experience a gradual decline of its population, from over 730 million now to under 700 million in 2050. Fertility rates in Europe are currently at about 1.5 children per woman, and, since this is below the rate needed to maintain a population (not considering migration), each European generation is producing about three-fourths of what they need to keep their population from declining. Similar social patterns are happening in the United States and Japan. There is a strong correlation between affluence and fertility rate. The question is, which will ultimately win? Can we increase global affluence fast enough to head off the long-anticipated global population crisis?

In many parts of the developing world, population is still increasing rapidly. One reason for this is the sociological inertia of many traditional cultures.

Before the advent of modern medicine, a typical married couple might have six or more children to ensure that two might survive to become adults. Modern medicine has drastically reduced infant and child mortality, but adherence to traditional family values in these regions still results in large families, especially among the rural poor. Very draconian measures, such as those practiced by the Chinese authorities, are implemented in many of the less developed countries to reduce population growth. The social consequences of China's "one child per family" laws are only now being revealed. With many families preferring boys over girls, and with abortion being legal, it is thought by many that the current male-to-female population imbalance in China can be attributed to the selective aborting of female babies. Is this a moral choice we must make for the good of the planet? (The authors certainly hope not.)

Must Population Ultimately Crash?

A peak global population of about 10 billion people around mid-century is a likelihood. But what happens next?

One person who contended with ultimate limits to population growth was Thomas Malthus, a late eighteenth-century British economist. According to Malthus, agricultural output generally increases arithmetically. For example, a farmer might have 10 acres under cultivation in 2010. He might add 1 acre per year so that he has 15 acres under cultivation in 2015.

However, population increase in Malthus's era was geometrical. To gain an understanding of geometrical increase, consider the fable of two men competing in a chess game. The winner requests that his prize should consist of one grain of rice on the first square of the chessboard, doubling the number of grains on each square thereafter: two on the second square, four on the third square, and so on. Although it might seem that the losing player had gotten off cheaply, the progression results in far more than the world's annual rice production well before the loser gets to the final, 64th square of the chess board.

Even though Malthus used these arguments to predict that agriculture would ultimately fail to supply ample foodstuff and the population would decrease due to famine, such a dire future is not inevitable. Population growth does seem to be slowing as couples in developed countries have fewer children. Agricultural science has thus far proven effective in developing more efficient crop strains.

However, there are more subtle forces to contend with. For instance, the expanding human population results in habitat destruction for other large mammals. As various mammal species face population reductions, the parasites infesting these organisms can mutate to change hosts. Since a numerous host is the human species, medical science faces a continual challenge that will only get worse as biodiversity decreases.

Space Technology and Human Population Increase

Since the subject of this book is the application of Solar-System resources to maintain and improve Earth's environment, it is worth considering what space technology can do to alleviate the population explosion. Bur first let's state what it cannot do. Wholesale migration of excess humans to cosmic sites will almost certainly not alleviate the population crunch.

Assume for the sake of argument that we eventually learn how to construct deep-space habitats for large numbers of humans using resources found on the Moon or nearby asteroids. (This concept, after being proposed by Princeton University physics professor Gerard K. O'Neill in the 1970s, was then developed theoretically by numerous researchers.) If population grows by 100 million people per year, about 300,000 people would need to depart Earth and travel into space each day.

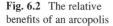

Fig. 6.2 The relative benefits of an arcopolis

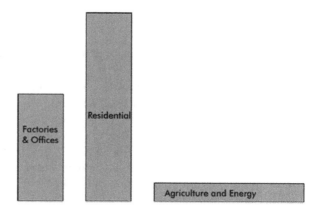

At present, if you wish to spend a week as a tourist aboard the International Space Station, the bill from the Russian launch services will amount to the equivalent of about $20 million. Let's say that breakthroughs and technological improvements eventually reduce this to $1 million, including baggage, life support, and other essentials to support the space immigrants and that the cost of producing the space habitats is reduced to zero, probably by application of a robotic labor force. The global space immigration cost per day would still amount to about $300 billion—a somewhat daunting figure. The entire global economy—and more—would be devoted to the space immigration effort.

Furthermore, such an approach might fail to have the desired effect. The large-scale European emigration to the United States in the nineteenth and early twentieth centuries did not substantially reduce the populations of European countries. New babies were rapidly produced to fill the gaps left by the emigrants.

However, space-age technology, such as closed-loop life support, recycling, efficient housing concepts, and intensive agriculture, can also be applied on Earth. In the late 1960s, architect Paolo Soleri introduced the concept of arcology. An "arcopolis" is an Earth-bound version of a space habitat or interstellar ark, used to house vast numbers of people in a compact city-state. Such a complex is represented schematically in Fig. 6.2.

Let's consider the effectiveness of such a compact, highly urbanized lifestyle on an overpopulated Planet Earth. Assume first that each of the 10 billion people on the planet in 2040 requires about 300 square feet (33 square meters) of comfortable living space. If everybody lived in a one-story house, the human-habitation "footprint" would have an area of about 3 trillion square feet, or a million square miles, about 2 % of Earth's land surface area. When we factor in the amount of land space that must be devoted to agriculture, energy-production, transportation, and so on, there is not much room left for rainforests and parks.

Now consider instead that people choose to live in high-rise skyscrapers. Imagine a residential structure 1,500 ft (500 m) tall, a bit shorter than the world's tallest building. There are 150 stories in this hypothetical building. The building's footprint is 300 × 60 ft. Each story has enough floor space to comfortably accommodate 60 people, so the entire structure can house 9,000 people.

From the NASA-funded follow-up work on O'Neill's space habitat ideas, the closed ecological system agricultural space required in space to feed one person is about 600 square feet (about 60 square meters). Since Earth-based agriculture must contend with the day-night cycle and weather and seasonal changes, we will assume that 2,000 square feet are required to feed each person. As in space, the assumption is made that water is efficiently recycled.

Our 9,000-person arcopolis therefore requires a footprint of 18,000 square feet for living space and 18 million square feet (less than 1 square mile) for hydroponic agriculture. In the unlikely event that an entire global population lived in such structures, the total agricultural and residential space required to house and feed a population of 10 billion people would approximate 1 million square miles, which is about 2 % of Earth's land surface area.

Since an arcopolis would be compact, the energy-inefficient automobile would not be widely used for commuting. Energy-efficient, high-speed rail networks could be used to connect urban centers, as is already the case in Western Europe.

Efficient applications of available high technology, closed-environment organic farming techniques, recycling, and other technologies have the potential to allow comfortable living for a global civilization of 10 billion people. There will be plenty of room for parks, museums, theaters, sports stadiums, schools, hospitals, universities, and wilderness.

The reader might be happy to learn that Soleri's concepts represent an ongoing project rather than a dream for the far future. At the age of 88, Soleri works with the nonprofit Cosanti Foundation, which is developing a prototype ecologically sustainable community in the Arizona desert near Cordes Junction, about 65 miles north of Phoenix. As of June 2008, about 50 people live in this prototype "Arcosanti," which will house more than 1,000 upon completion. Workshops, seminars, and conferences are conducted at the site, which is ideally suited to ultimately export wind or solar energy.

Located on marginal land for agriculture or conventional human development, Arcosanti is partially underground. Equipped with solar greenhouses on 25 acres of a 4,000-acre preserve, it may serve as an urban laboratory for future experiments in high-density urban living.

Because construction and operation of this desert prototype arcopolis is not funded by government grants, the Cosanti Foundation has developed an innovative fund-raising strategy. About 65 miles south of Arcosanti, in Scottsdale, Arizona, tourists can visit Cosanti, which has been designated an Arizona historic site. Here, they can support the project (which may ultimately require $200 million) by purchasing bronze or ceramic Soleri wind chimes, which can easily be mounted in city or suburban backyards or courtyards and in rural sites.

The problems of implementing an arcopolis and related technologies over the next century are daunting but not insurmountable. The choice is collectively ours, but if we choose and plan wisely, the long-term future can be bright.

Further Reading

Many environmental references discuss population-related issues. Two of them are G. Tyler Miller, Jr., *Environmental Science,* 4th edition (Belmont, CA: Wadsworth, 1992) and Bernard J. Nebel and Richard T. Wright, *Environmental Science,* 4th edition (Englewood Cliffs, NJ: Prentice Hall, 1993). The first of these books introduces the J-shaped curve and presents the fable of the chess players.

Space habitat references include Gerard K. O'Neill, *The High Frontier* (New York: Morrow, 1977) and R. D. Johnson and C. Holbrow, editors, *Space Settlements: a Design Study,* NASA SP-413 (Washington, DC: NASA, 1977).

Paolo Soleri first published his arcology concepts in P. Soleri, *The City in the Image of Man* (Cambridge, MA: MIT Press, 1969). Updated material on this topic can be found at the following website: www.arcology.com. To learn more about the ongoing work of the Cosanti Foundation, you can visit it online at www.cosanti.com.

Chapter 7
Climate Change

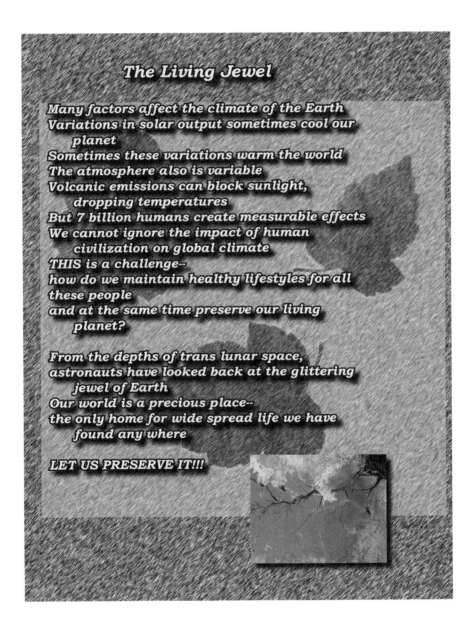

The Living Jewel

Many factors affect the climate of the Earth
Variations in solar output sometimes cool our
    planet
Sometimes these variations warm the world
The atmosphere also is variable
Volcanic emissions can block sunlight,
    dropping temperatures
But 7 billion humans create measurable effects
We cannot ignore the impact of human
    civilization on global climate
THIS is a challenge--
how do we maintain healthy lifestyles for all
these people
and at the same time preserve our living
    planet?

From the depths of trans lunar space,
astronauts have looked back at the glittering
    jewel of Earth
Our world is a precious place--
the only home for wide spread life we have
    found any where

LET US PRESERVE IT!!!

G. Matloff et al., *Harvesting Space for a Greener Earth*,
DOI 10.1007/978-1-4614-9426-3_7, © Springer Science+Business Media New York 2014

"Some say the world will end in fire,
Some say in ice,
From what I've tasted of desire
I hold with those who favor fire."
—From the poem "Fire and Ice" by Robert Frost

Here are some recent headlines from articles about global warming:

"Climate Change Promises Tough Times for Asia and Africa—World Bank Reports" *Inter Press Service*, June 21, 2013

"Map Shows Vast Regions of Ocean are Warmer," *Scientific American*, April 7, 2013

"The 10 warmest years in the record all occurred since 1998," *Global Temperature Update Through 2012, NASA*, January 15, 2013

"Arctic Melting Fast; May Swamp U. S. Coasts by 2099," *National Geographic News*, November 9, 2004

"Study: Earth 'Likely' Hottest in 2,000 years, "*The Associated Press*, June 22, 2006

"Unprecedented Warming Drives Dramatic Ecosystem Shifts in North Atlantic," *Science Daily*, November 7, 2008

These articles are alarming. They warn that Earth is getting warmer, and changes in the global climate are inevitable. The evidence of a global change in climate is mounting, and the consensus is that humans are responsible through our profligate emission of so-called greenhouse gases.

The authors of this book are not climatologists and are therefore not qualified to assert whether or not humanity is responsible for climate change. But as scientists, we can say that the body of evidence being put forth by climatologists asserting climate change seems credible. The theory that the change is being caused by human activity is almost as compelling and merits serious consideration and concern.

Before we go further, we should describe what we mean by climate change and review some recent history.

What Is Climate, and Why Is Climate Stability Important?

According to Wikipedia, climate "encompasses the temperatures, humidity, rainfall, atmospheric particle count, and numerous other meteorological factors in a given region over long periods of time, as opposed to the term weather, which refers to current activity." So a rainy day in London is an example of weather while a propensity for rainy days in London might describe an element of its climate. Obviously, different geographic regions have different climates. The climate of

Saudi Arabia is certainly different from that of Greenland and different still from that of Ecuador.

Climate is a driving force in determining where people live, their occupations, and often the relative affluence of those who live in any given location. Obviously, farmers in the American Midwest have a climate more favorable for agriculture than would be found in the African Sahara. Less obvious would be the connection between the rise and fall of entire civilizations wrought by changes in the climate. A report in the November 6, 2008, issue of *Science* discusses how the strength and frequency of Asian monsoons correlates with the rise and fall of the Chinese Tang, Yuan, and Ming dynasties within the last 1,800 years. It seems that during strong monsoon periods, rain-dependent rice crops flourished, and food was plentiful. When the monsoons became less frequent for many years (a climate shift), a correlation was observed between food scarcity and the fall of dynasties.[1]

According to Professor Gerald Haug of the University of Potsdam, Germany, and his colleagues, "Climate change is to blame for one of the most catastrophic collapses in human history." The collapse he was referring to was that of the Mayan civilization in Mesoamerica that ruled much of that region through about A.D. 800. Then, after being afflicted with a succession of lengthy droughts, their empire with a population of 15 million people collapsed. The droughts, combined with deforestation and the resulting soil erosion, ultimately appears to have devastated the Mayan civilization.[2]

Will the climate change we are now experiencing change the political, economic, and military balance of power in our modern world?

Changing Climate Is Not New

Based on the evidence left behind, we know that Earth has gone through at least three or four ice ages with advancing then retreating glaciers. From valleys carved through mountain ranges to rock walls deeply scarred by moving mountains of ice, the unmistakable signs of a much colder Earth with more snow and ice are visible today. In addition, analysis of ice cores containing tiny bubbles of ancient atmospheres reveal signature characteristics of multiple cold, glacial periods followed by warmer and more temperate ones.

Earth's climate is a very complex system with many variables, which makes it very difficult to model. It is also difficult to determine the specific effects of changes in many input parameters with any certainty. There are some variables, however, that are so significant that they can easily dwarf all other factors and produce dramatic changes in the global climate. Examples of these significant factors include changes in the Sun's intensity, volcanic eruptions, and changes in Earth's orbit.

Solar Variations

The Sun may appear to be constant, but it is far from unchanging. Within its volume it could contain over 1 million Earths. The outer surface of the Sun is about 6,000 K (10,000 °F), and, by nuclear fusion in its core, it converts approximately 5 million tons of matter into energy *every second*. This energy travels through space at the speed of light, and a fraction of it impacts Earth, providing the heat and light that sustain our planet.

There are sometimes explosions in the Sun's atmosphere, called solar flares, that release enormous amounts of energy into space, often impinging upon Earth. The frequency with which these flares occur varies with the solar cycle. During the peak of the cycle, several flares may erupt in a single day. During solar minimum, there might be less than one per week.

Despite these vagaries, Earth, on average, receives about 1.4 kW of energy per square meter. If measured from space and integrated over all wavelengths, the energy striking the atmosphere of Earth varies by less than one tenth of 1 % over an 11-year solar cycle. This continuous and mostly unchanging solar energy input into Earth's biosphere is called the solar constant.

As common sense might indicate, if the energy reaching Earth from the Sun increases, the average global temperature will increase. If the energy reaching us decreases, the average global temperature will decrease. Within fairly recent recorded history there was a dramatic decrease in Earth's temperature that was attributed to a decrease in the solar constant. Between the 1400s and 1700s, solar activity appeared to be minimal, and many infer that the energy reaching Earth decreased. There are widespread accounts of the "Little Ice Age" that resulted.

European historical records show that during this period the Thames, Bosphorus, and other rivers froze, as did New York's harbor. Northern Europe and North America experienced much colder summers, with commensurately shortened growing seasons. All over the world there were reports of glaciers expanding and record snowfalls.

Some historians surmise that the much colder climate resulted in the demise of the Norse settlements in Greenland, paving the way for other Europeans to rediscover North America and make it their own.[3] One has to wonder how the history of the last 700 years might have been different if the Sun had not gone relatively inactive, and these settlements had prospered and grown. There might not have been a need for Columbus and his Spanish-funded voyages across the Atlantic.

Volcanoes

Volcanoes emit tons of ash and aerosols during eruptions. Large eruptions spew forth larger amounts of both, changing not only the local environmental conditions but also climate worldwide. These dust and aerosols reflect an additional fraction of the sunlight falling upon Earth back into space, resulting in less energy actually striking the planet.

One of the best-documented examples of this occurred in 1815 with the eruption of the Tambora Volcano in Indonesia. Some historical accounts record that the years following the volcano's eruption were up to 5 ° F cooler than normal, producing at least 1 year "without a summer." Needless to say, it is difficult to provide the population with enough to eat if a growing season is lost.

Orbital Variations

Over very long periods of time, small variations in Earth's orbit as it goes around the Sun result in Earth being slightly further away from the Sun and therefore receiving less energy from it. During those periods, the summers would not be as warm and the winter snows would not completely melt away. As anyone who has been outside on a sunny day following a snowstorm can tell you, it is blinding. The sunlight reflects from the white snow, leaving little of its energy behind. With more of the warming sunlight reflected, Earth grows cooler still, resulting in yet more snow accumulation. This "positive-feedback" cycle repeated over many years results in an ice age. When the orbital variations change still more, and if the amount of sunlight increases, then the result will be glacial melting and warmer global temperatures.

Climate Is Changing: Rapidly

Since 1850, and as of this writing, 2005 was one of the two warmest years on record, followed by 1998, 2002, 2003, and 2004. The Arctic Ocean, once covered with massive ice sheets year round, is melting at an alarming rate, as are glaciers in Greenland (Fig. **7.1**). According to satellite observations and measurements from shipping and aircraft records, the Arctic ice is well below its average level and is dropping fast.[4] Observations from the other side of the world show that the Antarctic ice sheet is gaining mass and getting larger.

Areas affected by sustained drought are growing as rainfall patterns change (see also Chap. 13). Snowfall in European and American mountain ranges is in decline. The months that bring rain to the American Southwest are shifting from October through April to October through March, resulting in a loss of 1 month's rainfall and increasing the length and severity of that region's fire season. These are but a few examples of how Earth's water cycle is changing before our eyes.

The onset of the greening of spring, when plants begin to sprout new growth, comes earlier almost every year. Farmers are adapting to the change and planting their crops weeks earlier than in the past. The scientific and anecdotal evidence for climate change mounts.

Fig. 7.1 Meltwater creates an 18-m deep canyon in a polar ice sheet (image courtesy of Ian Joughin)

Are People to Blame?

There is evidence that human activity is responsible for the current period of changing climate. The prevailing theory is that our emission of greenhouse gases such as carbon dioxide is resulting in more heat being trapped by the atmosphere, thus increasing average global temperatures. According to the Intergovernmental Panel on Climate Change (IPCC) in 2007, the atmospheric concentration of carbon dioxide in 2005 was 379 parts per million compared to the preindustrial levels of 280 parts per million.

In an attempt to get a somewhat pristine measurement of atmospheric carbon dioxide levels (away from direct contributions from local human activities), scientists have been using instruments at the Mauna Loa Observatory in Hawaii. Figure 7.2 shows even more recent data from the Mauna Loa site. According to the data, the level of CO_2 is increasing every year and will shortly rise above 400 parts per million. Numerous studies indicate that this increase correlates with a rise in measured temperatures rather convincingly.

If this is a correlation, meaning that the two variables of temperature and atmospheric carbon dioxide levels just by happenstance change at the same rates and at the same times, then we have nothing to worry about. But if there is a causal connection, meaning that one (the level of carbon dioxide in the atmosphere) results in the other (the atmospheric temperature), then we humans had better take notice, as we are entering a period in which carbon dioxide in the atmosphere is increasing at an unprecedented rate.

Fig. 7.2 The *varying curve* shows the mole fraction of carbon dioxide in the atmosphere as measured by NOAA in Hawaii. The *solid line* is the average (image courtesy of NOAA and Scripps Institution of Oceanography)

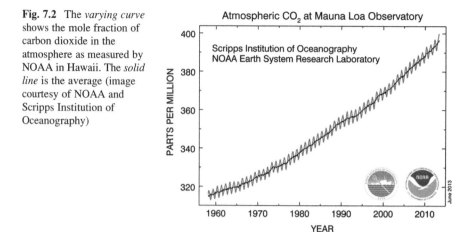

What, then, will be the ultimate impact on global temperatures? How will the climate change in response? The answer is that we do not know, but scientists have models that make some dire predictions.

As water warms, it expands and takes up more volume. Since Earth is a water world, with approximately 70 % of Earth's surface covered in water, even a modest increase in sea temperature will result in the sea level rising. If you add the water from melting Arctic glaciers, then it will rise even more. The consequences of a dramatic rise in sea level will potentially include the destruction of coastal wetlands and barrier islands (making our coasts more susceptible to damage from tropical storms), and a greater risk of flooding in coastal communities. Some islands may become uninhabitable, resulting in the displacement of the people living on them.

If the climate warms, diseases once restricted to the tropics will spread to more temperate climes, infecting people previously not threatened by them. Diseases such as malaria and dengue fever are among those now observed to be spreading beyond their normal habitats. In addition, warmer temperatures can cause higher incidences of food-borne illnesses; the warmer temperature causes food to spoil more rapidly and some bacteria to reproduce more vigorously.

Studies showing the impact of increased temperatures on food crops yield similar bad news. A widely reported study by Lawrence Livermore National Laboratory and Stanford University estimates that the annual yield of wheat, rice, corn, soybeans, barley, and sorghum will decrease by 3–5 % for every 1° of temperature increase.[5] Given that these crops are a significant source of food for much of the world, one can only surmise what might result, physically and politically, if these estimates are accurate.

Other species on Planet Earth will also be affected by the increase in temperature. For example, when carbon dioxide mixes with water, it forms carbonic acid. As we pump more and more carbon dioxide into the atmosphere, more and more of it will interact with Earth's ocean water, forming carbonic acid as a result.

The increased acidity of seawater damages ocean species with calcium carbonate shells as well as the coral reefs. Some scientists predict that up to 97 % of the world's coral reefs could be destroyed if the global temperature rises just 3.6 °F.

Some impacts of a warming Earth are counterintuitive. Take, for example, the case of the extreme Northern Hemisphere getting warmer. One would expect that warm-water fish of the middle Atlantic would be migrating northward as their ecosystem expands. But the opposite appears to be happening: fish normally found only in the cold waters of the north Atlantic are now being found further south than ever before. Why? Because melting Arctic ice sheets and glaciers are releasing cold water into the Atlantic, which is then carried southward by the ocean's currents. This colder water expands the ecosystem of the northern Atlantic fish, allowing them to venture further south than normal.

A review of the literature and the popular press will return more estimates, scientific or otherwise, postulating the negative (and some positive) effects of climate change. Unfortunately, there is a lot of hype surrounding the subject, with advocates of certain political views seizing on the issue of climate change to advance their agendas and potentially exaggerating the impact of climate change to suit their own ends. The converse is also happening. Those who do not want to make changes in the way they live or do business will take advantage of any scientific study questioning the reality of any aspect of climate change and attempt to use it to say that the threat is not real and that we should continue business as usual.

The authors of this book take a more cautious view. Human beings can and do affect the environment and, potentially, the climate. If we can make changes in how we live and conduct business so that we are better stewards of Earth, then we should seriously consider doing so. Later chapters describe what space development can contribute toward minimizing the negative environmental impact of our advanced technological society. We cannot afford to ignore the risks, nor can we afford to ignore a path that might eliminate those risks.

References

1. "Climate Change: Chinese Cave Speaks of a Fickle Sun Bringing Down Ancient Dynasties," Richard A. Kerr, *Science,* 7 November 2008, 322: 837-838.
2. "Climate and the Collapse of Maya Civilization," Gerald H. Haug, Detlef Günther, Larry C. Peterson, Daniel M. Sigman, Konrad A. Hughen, and Beat Aeschlimann, *Science,* 14 March 2003, 299: 1731-1735.
3. "Interdisciplinary Investigations of the End of the Norse Western Settlement in Greenland," L. K. Barlow, J. P. Sadler, A. E. J. Ogilvie, P. C. Buckland, T. Amorosi, J. H. Ingimundarson, P. Skidmore, A. J. Dugmore and T. H. McGovern, *The Holocene,* 1997, 7: 489–499.
4. "Satellite Evidence for an Arctic Sea Ice Cover in Transformation," Ola M. Johannessen, Elena V. Shalina, Martin W. Miles, *Science,* 3 December 1999, 286 (5446) 1937–1939.
5. "Global Scale Climate-Crop Yield Relationships and the Impacts of Recent Warming," David B. Lobell and Christopher B. Field, 2007, *Environ. Res. Lett.* **2,** 014002.

Further Reading

To learn more about the effects on climate from the variations in Earth's orbit around the Sun, see
   Milutin Milankovitch's seminal 1930 paper on the topic, "Mathematical Climatology and the
   Astronomical Theory of Climate Change." More information about Milankovitch is available
   in J. D. Macdougall's book, *Frozen Earth: The Once and Future Story of Ice Ages* (Berkeley,
   CA: University of California Press, 2004).
A comprehensive listing of the evidence supporting the reality of contemporary climate change
   can be found in K. E. Trenberth, P. D. Jones, P. Ambenje, et al., "Observations: Surface and
   Atmospheric Climate Change," in S. Solomon, D. Qin, M. Manning, et al., eds. *Climate
   Change 2007: The Physical Science Basis. Contribution of Working Group I to the Fourth
   Assessment Report of the Intergovernmental Panel on Climate Change* (New York: Cambridge
   University Press, 2007).

Chapter 8
Vanishing Life

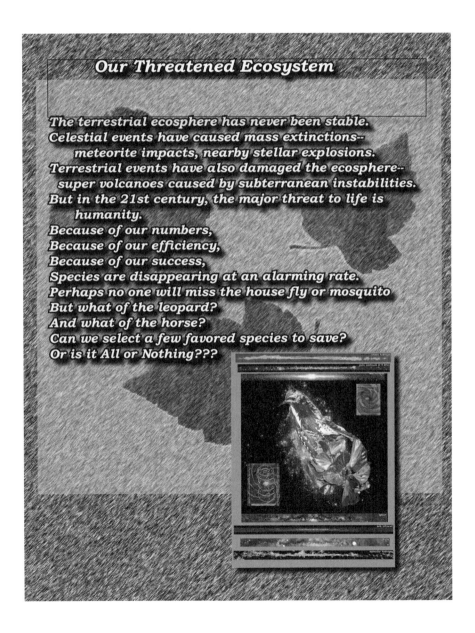

Our Threatened Ecosystem

The terrestrial ecosphere has never been stable.
Celestial events have caused mass extinctions--
  meteorite impacts, nearby stellar explosions.
Terrestrial events have also damaged the ecosphere--
  super volcanoes caused by subterranean instabilities.
But in the 21st century, the major threat to life is
  humanity.
Because of our numbers,
Because of our efficiency,
Because of our success,
Species are disappearing at an alarming rate.
Perhaps no one will miss the house fly or mosquito
But what of the leopard?
And what of the horse?
Can we select a few favored species to save?
Or is it All or Nothing???

G. Matloff et al., *Harvesting Space for a Greener Earth*,
DOI 10.1007/978-1-4614-9426-3_8, © Springer Science+Business Media New York 2014

"Roots and leaves themselves alone are these,
Scents brought to men and women from the wild woods
and pond side,
Breast-sorrel and pinks of love, fingers that wind
around tighter than vines,
Gushes from the throats of birds hid in the foliage
Of trees as the sun is risen."
—From the poem "Roots and Leaves Themselves Alone" by
Walt Whitman

When people think of the issue of decreasing biodiversity, they may think of pets, domestic animals, majestic beasts, and useful plants. Although many beasts and plants are threatened (pets and domestic animals are not threatened), the problem runs far deeper.

There is an interconnected web of life, and it is very difficult to cull one species without affecting others in the ecosystem. To further complicate things, there is not one ecosystem but many. If we damage one, no one knows what the consequences may be for others.

We should feel far from hopeless, however, about our ability to lessen humanity's harmful effects on the ecosystem. Coincidentally, our understanding of the ecosystem is growing as rapidly as our ability to alter it. More and more people are coming to view humanity as part of life's fabric, not as something outside of nature. This understanding may lead to the enlightened knowledge required to save many species great and small.

The Origins of Biodiversity

We know that there are many different species. But has this always been true? Will this always be true? Sadly, there is no real answer.

To investigate the origins of biodiversity, it is necessary to consider the origins of biological life forms. Scientists generally assume that life will originate naturally from nonliving systems given the proper conditions, such as appropriate temperature, moisture, and nutrient mix. But as long as we have only one example of a living world, we may never know if this, or the rival hypotheses that life's origin is an exceedingly improbable random event or the result of divine intervention, is correct.

Even if life is confirmed on Mars, Europa, or some other solar system world, the debate may not be resolved. It is not impossible that life could arise at only one location within a given solar system and be transferred by cosmic or geological events—meteorite impacts or volcanoes—to other planetary surfaces.

However, what we do know is that fossils of terrestrial life exist, thus indicating that life originated on our planet almost 4 billion years ago. Within a few hundred million years of this origin, early life had radiated to form the progenitors of the first animals, plants, fungi, protozoa, and bacteria.

Long before the rise of mammals, long before the first human hefted the first spear, Earth teemed with myriad life forms. Today, human activities threaten this biodiversity. But geological and cosmic events have also threatened life in our planet's long history.

Natural Mass Extinctions

It is fashionable in some circles to believe that pre-human life existed in an idyllic state. But even in the most peaceful natural state, there is fierce competition among species and members of the same species.

All life is programmed to do its best to pass on its genes to future generations. The organisms that succeed are those best adapted to their environment. Failing organisms may vanish or become extinct. Random mutations will alter the genetic structures of the survivors so that vacant ecological niches in the biosphere will be filled.

Nature seems peaceful because the evolutionary process works in slow motion. Life spans of individuals are measured in years, decades, or centuries. But many species survive for millions of years.

Sometimes, however, the rate of change is accelerated. At intervals of tens of millions of years, some action reshuffles the genetic deck. Periodically, events occur that rapidly transform the terrestrial environment on a global scale. These are the so-called mass extinctions. We know that impacting celestial objects contribute to at least some of them. About 65 million years ago, an asteroid or a comet about 10 km across slammed into what is now the Yucatan in Mexico. Large dinosaurs and many other organisms vanished in the aftermath of this event, to be replaced by mammals and the so-called feathered dinosaurs (more commonly called birds).

Since the Sun orbits the center of the Milky Way Galaxy, taking about 250 million years to complete one revolution, our Solar System sometimes approaches star-forming regions. Exploding stars (also called supernovae) within these clouds may then bathe our world in gamma rays and X-rays, irradiating and extinguishing many life forms.

Natural terrestrial activities also cause some mass extinctions. Super-volcanoes many times larger than Krakotoa or Vesuvius have the same effect as celestial impactors; they enshroud the entire planet in a long-lasting layer of high-altitude dust that blocks sunlight and causes a precipitous drop in surface temperature.

When a mass extinction occurs, as many as 95 % of terrestrial species may become extinct. It may be 100,000 years or so after the event that the surviving organisms finally rebuild a healthy, but greatly modified, global ecosystem.

Our planet seems to be undergoing a mass extinction event right now. But this one is far different from those in the fossil record. We cannot blame an impacting celestial visitor. No supernova has occurred within our galactic vicinity, at least for many millions of years, and we cannot pin the blame on natural terrestrial events such as super volcanoes. Only one terrestrial species seems to be responsible for the extinctions. It is currently, without much foresight or planning, eliminating entire ecosystems at a prodigious rate. The guilty party is none other than *Homo sapiens*.

Natural Ecosystems

To better understand humanity's destructive role and to alter it, it is necessary to gain some understanding of the ecosystem. From deep space, Earth appears to be an integrated living organism. But on closer inspection, it has many separate components. Naturalists have attempted to categorize the various ecosystems that make up our living Earth. Ecosystems are those distinct environments, with the organisms adapted to surviving there.

In his 1984 epic book *The Living Planet*, David Attenborough listed and described the major natural distinct but interconnected terrestrial ecosystems. These include the oceans, seashores, fresh water bodies, volcanic calderas, polar ice sheets and tundra, forests, jungles, grasslands, deserts, and the sky. Also, because civilization has been widespread on planet Earth for 5,000 years or more, we must include the semiartificial ecosystem, including those organisms that have adapted to live near human concentrations.

Since the publication of Attenborough's book, scientific ecology has advanced. It is now understood that there exists another ecosystem, possibly more significant and elaborate than the ones listed above: the subterranean world of Earth's crust.

The web of life within and between the ecosystems is of varying strength. Humans, through over-fishing, can damage the deep-ocean ecosystem and influence the shore as well. But even if we are foolish enough to attempt such a deed, we cannot destroy all life on this planet. Even if we managed to wipe out surface, ocean, lake, and aerial life, subterranean forms would ultimately recolonize the planet's surface, though the recolonization might take tens of millions of years.

The Human-Directed Ecosystem

At least on a small scale, humans have been modifying natural ecosystems for millennia. At the dawn of the Neolithic, around 10,000 years ago, our ancestors domesticated the dog and cat. Dogs were bred from the fiercest canines—the wolves—and became the hunting partners of humans. Cats, bred from equally fierce small felines, were useful in eliminating vermin inhabiting the grain depositories of early farmers. Grain itself—the very staff of life—has been altered from its natural state by so many generations of controlled genetic manipulation that it almost certainly would not survive well in the wild.

Many organisms in nature have co-evolved with other creatures. For instance, many species of hummingbirds have developed beaks specially equipped to pollinate certain selected species of orchards. And the orchards have expended a great deal of genetic energy making themselves more attractive to their pollinators. So if human activities eliminate a flower, certain bird species may vanish as well.

It should not be believed that all human-caused biodiversity degradation has been due to the actions of evil, selfish, or uncaring people. A laudable goal of human farmers is to grow more crops to feed a growing human population. So farmers have elected over the years to enlist the services of biochemists to develop insecticides such as DDT to protect crops from insect vermin and to control the mosquito population, carriers of the deadly malaria virus. But one reason that DDT was banned was the discovery that, as a side effect, it also weakened the eggshells of certain nautical birds.

Because we have considered human effects upon other species for only a short time, from an evolutionary viewpoint, human carelessness also has an impact. Large cities concentrate certain organisms—such as the rat, roach, and pigeon—at the expense of others. Border fences can disrupt the migratory patterns of many creatures.

The monarch butterfly conducts epic annual migrations from the northeastern United States to its winter home in Mexico. Ecotourism in the Mexican forests must be rigorously overseen, to make certain that hordes of well-meaning butterfly lovers do not put this species at risk.

Human energy-production systems, even of the most benign forms, can have serious effects upon local ecosystems, if they are designed without adequate foresight. Wind energy is thought of as one of the greenest electric-power production possibilities. But small mammals preyed upon by eagles have learned to hide beneath the spinning blades of California wind turbines, causing mass fatalities among the American national bird's population. Large wind-power farms may also change down-range wind patterns, which can affect the migration patterns of insects and other small organisms. A 2012 study published in the journal *Nature Climate Change* found that the temperature increased ~0.72° in wind farm regions compared to areas without turbines, causing a localized warming, particularly at night.

As developed and developing nations turn toward biofuels such as ethanol to supply transportation fuel and reduce greenhouse-gas emissions, unforeseen effects on the ecosystem can occur. Most biodiversity on Planet Earth is found in the rain forest, which also serves as an absorber of carbon dioxide and producer of oxygen. If developing equatorial countries destroy their rain forests to produce biofuel exports, both biodiversity and global climate will suffer.

Even the most noxious, destructive organisms should not be eliminated from the global ecosystem, no matter what our opinions of them might be.[1] Consider, for

---

[1] The authors are willing, however, to concede that some species are best eliminated. We cannot think of a single positive environmental benefit of the smallpox virus.

example, the termite. Among homeowners reading this book, this wood-munching insect will find few admirers. But in nature, this creature eats and reduces the cellulose in dead wood. If we somehow managed to rid the planet of the termite, forest ecosystems would suffer.

The collapse of bee colonies has been a disturbing global phenomenon in the last few years. Bees play an important part of the food chain upon which all depend, and if they should vanish, which is a very low likelihood event, it would be disastrous. Since the current die-off began in the 1990s, there have been reports of between 30 and 90 % death rate in some commercial bee colonies. As of this writing, the cause of the die off is unknown and may yet be found to be unrelated to human activity; Some scientists attribute the decline to a variety of pesticides recently put into use.

According to Zakri Hamid, a member of the United Nations Intergovernmental Platform on Biodiversity and Ecosystem Services (IPBES), the increasing loss of biodiversity poses a "fundamental threat" to the "survival of humankind." Why? The threat is in large part due to the decreased genetic diversity of our food crops, making them more susceptible to disease. For example, the variety of apples grown in the United States is down from hundreds at the turn of the twentieth century to about ten today.

The rain forests are the most vulnerable of the major terrestrial ecosystems, largely because the depth of fertile soil is not large. But these regions are also home to the most diverse population of species. Although our catalog of rain forest species is incomplete, many rain forest organisms are known to have significant medicinal properties. It is alarming that unconstrained commerce is destroying rain forest environments and species before we even understand the resources of these regions.

What Can We Do?

The point of this book is not to predict doomsday, but to help prevent it. If we approach these problems in an enlightened manner, they can be solved. First, we might consider the symptoms. Nations might make better use of Earth-resource monitoring technology. Since the problem of ecosphere degradation is global, international data sharing and cooperation are essential. The same space-based technology used to monitor terrorist threats can be applied to protect endangered terrestrial species.

Responsible eco-tourism has already emerged as a positive influence on biosphere protection. Let us hope that it will be expanded during the critical decades ahead.

As we replace and supplement fossil fuels, planning on an international level is required. Biofuel use, wind turbines, and other "green" technologies should be implemented with as much environmental planning as possible. Space research on

Fig. 8.1 An artist's concept
of Earth as an integrated
organism, connected by
DNA (courtesy of NASA)

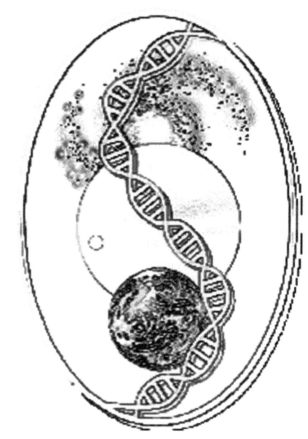

closed ecological systems may lead to biofuel production modes that have only transient effects upon the rain forest and other threatened terrestrial ecosystems.

However, all these approaches treat symptoms. To attack the root causes of the human-caused mass extinction currently under way, human attitudes toward our species and our planet must change. Many consider the best hope to be a widespread adoption of the Gaia hypothesis pioneered by James Lovelock and Lynn Margolis—or at least some aspects of it. Instead of viewing ourselves as separate, disconnected organisms, we might profitably consider each individual life form to be analogous to a living cell and each species to play the role of an organ in an integrated, living planet. Earth might be considered as a living entity, with all of her components linked by the shared DNA molecule (Fig. 8.1).

Humans collectively might be considered as the nervous system of this planetary organism. Let us hope that we wisely use our intelligence to nurture and preserve life on our small planet.

From the Headlines

Scientists warn Earth's entire biosphere nearing a catastrophic "tipping point"
The Christian Science Monitor, 2012.
One-fifth of all invertebrate species facing extinction: "Extinction of humans could soon follow"
Zoological Society of London, 2012.
Oceans on the brink of catastrophe
The Independent, 2011.
Half Of All Species May Be Extinct In Our Lifetime
Proceedings of National Academy of Sciences, 2008.
Populations of all wild animals crashing—down 30 % since 1960.
Canadian Broadcasting Corporation, 2008.
Mass extinction taking place in Earth's oceans
Christian Science Monitor, 2010.
12,000 Threatened Species: "Only Scratching the Surface"– 2003 Red List Released
U.K. Guardian, 2003.
All coral reef species face extinction by the end of this century
CBS News, 2010.
Rising Acidity of World's Oceans May Cause Marine Mass Extinction
London Times, 2006.

Further Reading

An incomparable work on terrestrial biodiversity is David Attenborough, *The Living Planet* (Boston: Little, Brown, 1984). A good reference on the Gaia hypothesis is James E. Lovelock, *The Ages of Gaia* (New York: Oxford University Press, 1988)

Chapter 9
Diminishing Energy

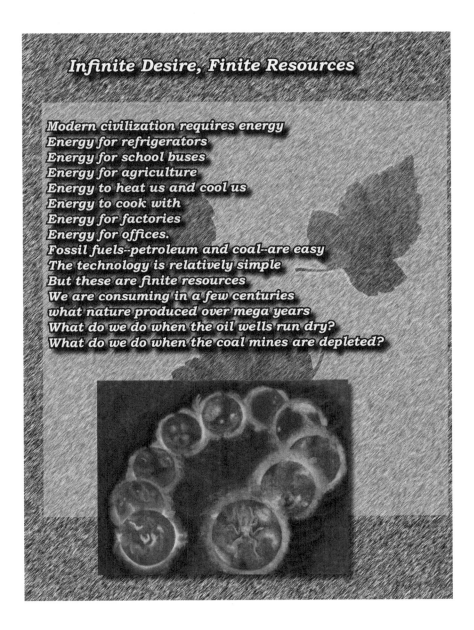

Infinite Desire, Finite Resources

Modern civilization requires energy
Energy for refrigerators
Energy for school buses
Energy for agriculture
Energy to heat us and cool us
Energy to cook with
Energy for factories
Energy for offices.
Fossil fuels--petroleum and coal--are easy
The technology is relatively simple
But these are finite resources
We are consuming in a few centuries
what nature produced over mega years
What do we do when the oil wells run dry?
What do we do when the coal mines are depleted?

G. Matloff et al., *Harvesting Space for a Greener Earth*,
DOI 10.1007/978-1-4614-9426-3_9, © Springer Science+Business Media New York 2014

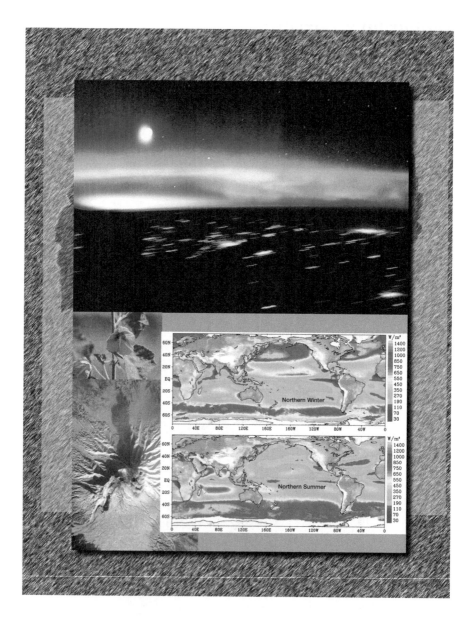

Northern Winter

Northern Summer

"They are all gone away,
The House is shut and still,
There is nothing more to say.
Through broken walls and gray
The winds blow bleak and shrill;
They are all gone away."
—From the poem "The House on the Hill" by Edwin
Arlington Robinson.

As mentioned in Chap. 4, the immutable laws of thermodynamics appear to place a limit on how well we can be stewards of our planet. To recap, the first law states that all energy is conserved. This means that no matter what you do, you cannot get more energy out of system than you put into it. And some of that energy is inevitably wasted, which is the second law of thermodynamics. Translated into a discussion of global energy supplies, at least when we limit ourselves to discussing the energy problem and planet Earth, these two laws tell us that no matter how clever we may be, no matter how resourceful we become, no matter how hard we try, we will always waste some of the energy we consume.

Where is energy consumed? In 2004, the United States consumed approximately 25 % of the world's energy. About a quarter of that energy was consumed for transportation, a third was for industrial use, and the remaining was fairly evenly split between residential and commercial users. Other countries use energy, too, and as the rapid industrialization of formerly low-energy consuming countries continues, their need for energy is rising. The reader should note what constitutes almost all of the projected growth is oil, coal, and natural gas (Fig. 9.1).

Fortunately for our civilization, nature took millions of years to convert the energy received from the Sun into plentiful and fairly compact energy storage systems. Oil, coal, and natural gas, the so-called "fossil fuels," are most simply understood as batteries, storing energy received by plants and animals from the Sun so long ago into a compact form that we have learned to use today. Over millions of years, fossil fuels are thought to have formed by the action of heat from Earth's core and pressure from rock and soil on the remains, or fossils, of dead plants and animals. Whenever you drive your car or turn on your cook stove, you are essentially burning energy from the Sun that fell upon Earth in the distant past and was stored until it was processed and piped to your car or home. Unfortunately, the world's supply of these resources is finite and we will eventually run out.

A recent debate has developed regarding the technology used to tap less easily mined reserves of one fossil fuel, natural gas. The advantage of natural gas is that it results in less greenhouse gas emissions than coal or petroleum. Since about 2000, a technique called hydraulic fracking has become more economical for obtaining this material from rock layers, especially shale.

Introduced in Texas in 1938 and later used in Russia and Europe, fracking works by using pressurized fluids to allow gas and petroleum to migrate along channels or fractures in the rock. Enormous pressures are required—up to 1,000 atm. A wide variety of fracturing fluids are utilized with chemical additives that increase the

Fig. 9.1 World marketed energy use by fuel type as projected by the U.S. Energy Information Administration (from U.S. Energy Information Administration [EIA], International Energy Annual 2005. Washington, DC. EIA, June–October 2007)

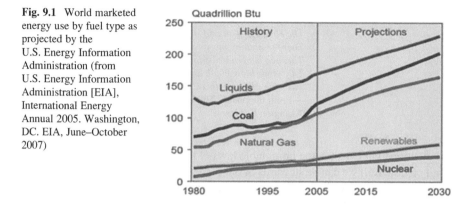

viscosity of the fracturing fluid. According to the International Energy Agency, as much as 7 trillion cubic feet of natural gas deposits may be obtained using fracking.

However, this technology has serious environmental consequences, which resulted in France's 2011 ban on hydraulic fracking. Disadvantages include risks of groundwater contamination, local air-quality degradation, blowouts and explosions, migration of gases and chemicals to the surface and associated health risks, and the possibility of bureaucratic issues that result in the mishandling of waste. Natural gas retrieved using fracking may also emit more methane and carbon dioxide when combusted than natural gas obtained using more conventional techniques. Some radioactivity from heavy metals flowing to the surface may be released at certain sites and local seismic effects might be produced.

Health effects have been noted in rural populations in New York State and Pennsylvania near fracking sites. So if fracking is to increase, it must be rigorously regulated to protect local and regional populations.

There is some debate as to when our reserves of oil, coal, and natural gas will be expended. The date when the well will run dry for each of these depends upon several factors: the rate at which they are used, the amount that remains to be used in known or yet-to-be discovered locations, and, of course, how serious we humans become about conserving the fuel for later generations. But one thing is certain. We will run out of each and every one of these fuels. The only open question is when. Whether it will be in 50, 100, or 300 years is a matter of mostly academic debate. But that date will eventually arrive and we'd better have a backup plan. What are the options?

Nuclear Power (Fission)

The United States, France, and Japan together produce approximately 57 % of the world's nuclear-generated electricity. In the United States, 19 % of the electricity consumed comes from nuclear power. In France, 78 % of electrical power is nuclear.

Fig. 9.2 Brown's Ferry nuclear power plant near Athens, Alabama, in the United States (photo is courtesy of the United State Nuclear Regulatory Commission)

Nuclear energy is not directly extracted from an atom and turned into electricity. Rather, the process by which power is generated from a nuclear source resembles how power is produced in a convention coal, oil, or natural gas fired plant—through the generation of heat. In this case, splitting uranium atoms in a process called nuclear fission produces the heat. When a specific target atom is struck by a neutron, it forms two or more smaller atoms releasing energy and, when properly configured, more neutrons. These neutrons strike other atoms, releasing yet more neutrons and more heat. The heat is used to boil water and produce steam, which then drives a turbine to make electricity (Fig. 9.2).

The fuel for the nuclear reactor is usually the element uranium. Uranium is fairly common, and traces of it can be found in most rocks, soils, and in the oceans. This uranium can be processed for use in nuclear power plants directly, or it can be used in another type of reactor that makes its own fuel. The latter is called a breeder reactor because during the nuclear chain reaction process it "breeds" fuel for future use. With known uranium reserves and breeder reactor technology, nuclear power could provide humanity with abundant electricity for millennia. It is also important to note that generating electricity in a nuclear power plant does not produce any greenhouse gases.

Nuclear power is not without its critics, however, and some of the concerns raised are serious. The safe operation and eventual disposal of nuclear power stations is of paramount importance. Even a small radiation leak is environmentally unacceptable, and a large leak could prove catastrophic. Fortunately, the accident at Three Mile Island in the United States killed no one and produced no environmental damage. The same cannot be said about Russia's Chernobyl accident. Chernobyl was by far the worst reactor accident in history, producing both loss of life and localized environmental damage. The accident resulted from the reactor operators ignoring their own rules, disabling key safety features that would have otherwise prevented the accident. Above all, the reactor had no containment vessel to trap the radioactive gases that were released—a required feature of virtually all commercial reactors used to generate electricity in the rest of the world.

The March 2011 nuclear tragedy in Japan, which released only about a tenth of the radiation released at Chernobyl, has far reaching consequences for advocates of nuclear power. The reactors at Fukushima had been designed to withstand the most severe recorded Japanese earthquakes and associated tsunamis. But the one that occurred in March of that year was off the scale. The most powerful earthquake to have hit Japan since accurate seismic record keeping began in about 1900, the epicenter of this event was about 70 km east of the Oshika Peninsula of Tohoku and about 32 km below the surface of the ocean. It is one of the five most powerful earthquakes since 1900, and it resulted in more than 16,000 deaths.

Because of the quake-produced tsunami that overwhelmed the region, electricity was off-line. The cooling systems therefore failed and resulted in meltdowns in the cores of several reactors in the Fukishima complex. Hydrogen gas build-up resulted in explosions rupturing the containment vessels and releasing radioactivity to the environment. A 20-km evacuation radius was required around the crippled reactors. The Tokyo government banned the sale of food grown in the area and local tap water was declared to be unsafe for infants.

There were no immediate deaths from the accident, but 6 workers exceeded the lifetime radiation exposure limit and 300 received significant radiation doses. Some sources estimate that radiation released at Fukishima may ultimately result in as many as 1,000 deaths.

In December of 2011, the Fukishima plants were declared to be stable. But decades may be required for decontamination. Many fission reactors may be closed worldwide as a result of this incident. Anti-nuclear sentiment has increased in many countries other than Japan, including Germany, India, Italy, Spain, Switzerland, and the United States.

Other than public distrust, the biggest drawback to nuclear power, in the authors' opinion, is the safe storage of the radioactive waste generated as the byproduct of a reactor's operation. The United States has no active central repository for this deadly waste, which can remain toxic for many thousands of years. Much of the generated waste is in "temporary storage" at the plants themselves, awaiting shipment to a more permanent storage facility elsewhere. Unfortunately, this temporary storage is looking more and more like a permanent situation and is certainly not a viable permanent solution. Susceptible to accident or deliberate tampering, this on-site stored waste is a relatively near-term problem that must be fixed.

Nuclear power plants can, in the wrong hands, be used to produce fuel for nuclear weapons. A serious risk of the technology is that its stated peaceful purposes can fairly easily be diverted to making bombs and increasing the risk of nuclear war occurring somewhere in the world. Having this technology in use by unstable or rogue states could prove to be a major danger.

Another drawback is the sheer scale at which nuclear reactors would have to be built in order to replace power plants currently using oil, coal, and natural gas. Thousands of new nuclear power plants will be required just to meet existing demand. The risk from diversion of nuclear fuel to terrorists, accident, or misuse of reactor technology by countries seeking to develop nuclear weapons make this a potential global energy solution that should only be undertaken when most, if not all, other options have been ruled out. The key word in this assertion is the word, "global."

Unquestionably, the United States, Europe, and Japan, as well as other stable and peaceful governments, could build redesigned nuclear power plants with improved safety features to meet our near-term energy needs (over the next 50 years), but they should only be considered an interim solution. And care must be taken not to repeat Chernobyl or Fukishima!

Nuclear Power (Fusion)

There is a joke among physicists that goes something like this; "Fusion is the power source of the future—and it always will be." (There is an alternative version: "Fusion power is 30 years away—and it always will be.") This appears to be the unfortunate reality of fusion energy research for the last 50 years, and, barring a breakthrough, the trend seems likely to continue.

What is nuclear fusion and why does it appear to be so hard to use it for power production?

The energy produced in the Sun comes from nuclear fusion. In the Sun, hydrogen atoms are being tightly compressed and fused together to form helium. This process has been ongoing for 4.5 billion years and will continue for at least a few billion more. Creating a fusion reaction in the laboratory isn't the problem; that is something scientists began doing since the middle of the last century. The problems are with control and something called "breakeven."

An uncontrolled nuclear fusion reaction is a bomb. The hydrogen bomb uses nuclear fusion to produce tremendous devastation and should not to be confused with the fission bombs dropped on Japan during World War II. A fusion bomb uses the energy produced in fission bomb to get the reaction started, whereupon an uncontrolled fission reaction ensues, releasing significantly more energy than is possible with a fission bomb alone. Even the most primitive fusion bombs produced about 500 times as much energy as a fission bomb.

The fusion reaction can be controlled, and this too has been routinely achieved in the laboratory. The problem is that, for now, it takes as much or more energy to create and sustain a fusion reaction than it does to produce it. And as long as you have to put more energy into fusion than you get out of fusion, fusion will not be a viable (net) energy-producing process. The point at which you are able to extract as much energy as you input into a fusion reaction is called "breakeven."

Finding a consistent, economically viable way to reproduce the conditions in the Sun to produce fusion has been the challenge, and the solution, at least at the current pace of research, may be just "30 years away." However, if the resources of the Solar System are brought to bear, specifically the vast stores of helium-3 embedded in the lunar regolith, then nuclear fusion may not be so far away after all. (See Chap. 11 for more information about helium-3.)

Hydropower

Another source of electricity that produces no direct greenhouse gases is hydro-electric power. This power source comes from the use of flowing water, typically produced when a dam is constructed across a river. Instead of steam, as is used in a fossil fuel or nuclear power plant, hydropower uses falling water to turn a turbine and produce electrical power. The process is relatively clean and efficient. However, it is not without environmental consequences. When a dam is constructed, a river is usually turned into a lake. The lake encompasses land that used to be available for other purposes—as someone's home, farmland, industry, or, at the very least, as a habitat for various land animals. All of these uses are precluded when the area is flooded to create the lake that then provides the water for generating hydropower (Fig. 9.3).

Many of the power-producing dams in the United States were built in the 1930s and 1940s, long before our current age of environmental awareness. The environmental impact of building a single dam, let alone the many thousands that would be required to make a dent in our dependence on fossil fuels, would be unacceptable. And there may not be enough suitable locations for building these dams, even if the local environmental consequences were considered to not play a major role. All the easy sites are taken.

The pros and cons of hydropower expansion can be observed by reviewing recent developments in China. On July 4, 2012, the world's largest electrical power facility built at an estimated cost of 22.5 billion in U. S. dollars, one that produces 22,500 MW of power was completed. Called the Three Gorges Dam and located near the town of Sandouping on the Yangtze River, the plus side of this fully functional facility is that it may increase shipping and reduce downstream flooding. There will be a large reduction in Chinese greenhouse gas emissions, and reforestation near the project may result in an increase of 6,000 km^2 in forest cover. But on the downside, the Three Gorges Dam has resulted in the displacement of an estimated 1.3 million people. Many significant archaeological and historical sites have been flooded. Some species of river dolphins are endangered; one may already have become extinct. Cranes and sturgeon in the Yangtze are also affected.

Fig. 9.3 The Lake Guntersville Dam and hydroelectric power plant near Guntersville, Alabama (courtesy of the Marshall County Alabama Convention and Visitors Bureau)

Another form of hydropower is tidal power generation. Harnessing the tides to turn turbines and generate electricity is technologically feasible and has been demonstrated. Unfortunately, tidal power stations can only produce power along the cycle of the tides (peaking every 6 h), which is not likely to track the human power demand cycle. There is also the open question of what these large underwater turbines will do to the coastal environment and its aquatic residents. Hydropower may provide part of a global solution, but at best only a small part.

Wind Power

In a fossil fuel or nuclear plant, steam is used to turn a turbine and generate electricity—so why not use wind power? The idea is not new, and several wind farms are now populated with multiple windmills and regularly generating electricity. Currently, about 1 % of the world's electrical power is produced using wind.

Wind energy is plentiful, renewable, clean and free of greenhouse gas emissions. There is no technical reason why many thousands of windmills cannot be built and added to the power generation infrastructure today. But there are technical reasons why they cannot be the solution to the world's energy needs (Fig. 9.4).

The first problem is one of wind availability. The wind is not constant, and it does not increase when the power demand is high. Is it practical to have a power

Fig. 9.4 The Buffalo Mountain, Tennessee, Wind Park is the largest such park in the southeastern United States (courtesy of the Tennessee Valley Authority)

grid that has to shut down, or perhaps just "brown out," when the winds shift? Also, the amount of power generated is highly dependent on the speed of the wind. In studying one windmill farm, engineers found that about half the total energy available was produced in about 15 % of their operating time. The rest of the time the power output was low, reducing the amount available to consumers.

The other major problems facing wind power are environmental. Some of the best locations in the world for wind power generation are right in the middle of protected lands or on the horizon of majestic landscapes where people are loathe to allow large manmade structures to mar the view. Do we want to cover our public lands with huge towers and power lines? Or the oceans just off the coast of the northeastern United States?

Another environmental problem is the scale of the land use required to generate large quantities of power. For example, a 200-MW wind farm requires approximately 20 km^2 of land. This is mainly due to the spacing needed between each windmill so as to reduce the power generation losses that result when they are too close together. Fossil fuel plants with much higher power output require significantly less land.

Wind power is another good idea for supplementing the energy supply, but it, too, is not a viable long-term solution.

Geothermal Energy

Another niche power production system uses Earth's natural geothermal energy. In various parts of the world, steam generated deep within Earth escapes to the surface. The resulting hot springs have been used for centuries for bathing, heating, and as cures for various maladies. They are also popular tourist spots. The Old Faithful geyser in Yellowstone National Park is perhaps the most well known (Fig. 9.5). Early in the twentieth century, the first experiments were conducted in using this natural form of steam to generate electricity—and it works. Unlike most other forms of renewable energy such as wind and solar, which are typically intermittent, geothermal can provide base load power.

Across the globe entrepreneurs are tapping into geothermal energy to build power plants and generate electricity in a manner that does not pollute the atmosphere. Unfortunately, the very geologic formations that make this power source possible also may place limitations on its overall viability. In many cases, outside water must be pumped into Earth's "hot spots" to make steam and produce power. The water is then pumped elsewhere to cool. This process sometimes makes the nearby soil unstable, producing cave-ins and landslides, and sometimes small earthquakes.

The aforementioned concerns will limit how widespread the technology may be used. Conventional geothermal energy is available in only a few locations (where Earth has provided access to its deeply produced, innermost heat), and only some of those are sufficiently stable so as to allow for the building large-scale power plants.

Like wind energy, geothermal energy is an excellent supplemental energy source, but it is not scalable to meet the needs of a global, high technology civilization.

Biofuel

A lot of attention has been paid to biofuels in the last few years. Basically a biofuel is one that is derived from recently deceased plants (as opposed to fossil fuels, whose energy is derived from ancient deceased plants). The most common biofuel is alcohol generated from corn or sugar cane. With the increasing price of oil, making gasoline with 10–20 % biofuel alcohol has become cost competitive.

There are serious questions about the overall energy efficiency of the process, however, given that the plants must be planted and harvested using machines, processed by machines, and then distributed (again, by machines) to where the users may be located. Taking into account the energy required in each of these steps prior to consumers filling up their tanks, some scientists doubt that the energy output exceeds the energy input. In fact, it is possible that more energy will be consumed in making a liter of biofuel than will be extracted from it by our cars and trucks.

There is also the problem that raising the huge amount of plants that will be required to sustain a viable biofuel infrastructure may pit the use of arable land in

Fig. 9.5 Geothermal energy may be sometimes readily accessible, as seen in this photograph of the Old Faithful geyser in Yellowstone National Park. But it is the very fact that geothermal energy is only available in a few places that make it a popular tourist attraction to begin with (courtesy of the U.S. Geological Survey)

cultivating biofuel against that same land being used to grow food. Do we really want to have food and fuel competing against one another? We think not, especially in a world that still sees a large fraction of its population going to bed hungry each night.

Conclusion

None of the alternative energy sources described above is a bad idea. None should be dismissed as being able to contribute to our global renewable and green energy strategy. And none of them individually, nor taken together, provide a viable long-term solution to the world's increasing energy needs. No, something new and different must be tried—like getting power from space. For more on space solar power, see Chap. 12.

Further Reading

For further information on the thermal effects of wind farms, consult: L. Zhou, Y. Tian, S. B. Roy, C. Thomcroft, L. F. Bosart and Y. Hu, "Impacts of Wind Farms on Land Surface Temperature," Nature Climate Change, Vol. 2, pp. 539–543 (2012).

Chapter 10
Humans Before the Industrial Age:
A Desirable Ecological Goal?

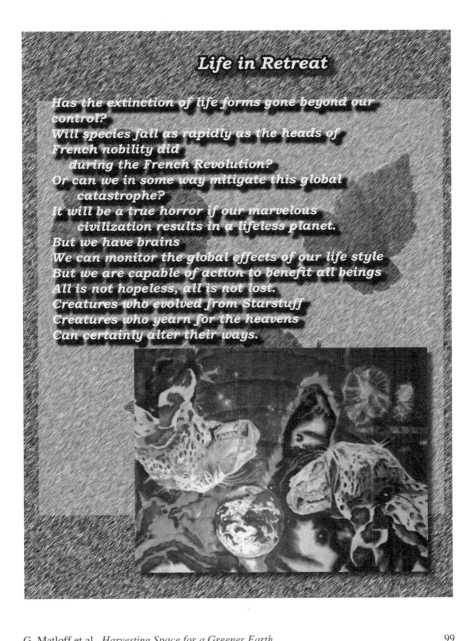

Life in Retreat

Has the extinction of life forms gone beyond our
control?
Will species fall as rapidly as the heads of
French nobility did
    during the French Revolution?
Or can we in some way mitigate this global
    catastrophe?
It will be a true horror if our marvelous
    civilization results in a lifeless planet.
But we have brains
We can monitor the global effects of our life style
But we are capable of action to benefit all beings
All is not hopeless, all is not lost.
Creatures who evolved from Starstuff
Creatures who yearn for the heavens
Can certainly alter their ways.

G. Matloff et al., *Harvesting Space for a Greener Earth*,
DOI 10.1007/978-1-4614-9426-3_10, © Springer Science+Business Media New York 2014

"We are the music-makers,
And we are the dreamers of dreams,
Wandering by lone sea-breakers,
And sitting by desolate streams;
World-losers and world-forsakers,
On whom the pale moon gleams:
Yet we are the movers and shakers
Of the world, forever, it seems."
　　　　—From the poem "Ode" by Arthur O'Shaughnessy.

All of us have experienced the darkness, felt the despair. Our civilization's problems seem beyond measure—an exploding population, pollution, energy shortages, climate change, terrorism, and nuclear proliferation are among these threats. There is nowhere in space to flee. Earth is filled up, the moon lacks air and liquid water, Mars is a lifeless desert, and the stars are just too distant.

However, if we cannot flee to greener pastures, perhaps we can maximize our chances of surviving a future global catastrophe. Then we can emerge from our holes and help direct the surviving human remnant on a different, simpler path. Some believe that if we forsake the world of the cities, cell phones, cars, and computers, then we can return to a simpler, sustainable pre-industrial golden age.

Has such an age ever existed? What would be the consequences if we actually retreated from our technologies?

The Paleolithic: Life in the Old Stone Age

Let's first imagine that we chuck it all. We reset the human clock as far back as possible and return to the good old days of the Paleolithic Era. The population was much smaller and human-caused pollution nonexistent in this, the longest period of human prehistory, which extends from about 2,000,000 to 10,000 B.C. We could all be movers and shakers then, since governments and bureaucratic structures did not yet exist. But our lives would not be very long, or very interesting.

Most human activity in this period centered on the quest for food. Since we did not farm or know how to maintain animal herds, it was necessary to gather nutritious herbs, follow game animals, and opportunistically steal the kills of larger predators.

This must have been a lot harder early in the Paleolithic, since fire had not yet been tamed and our best cutting tools were of hand-carved flint and rock. Until fire was mastered about one million years ago, our ancestors must have consumed their food raw. As well as lacking in taste, raw meat often contains various parasites. This fact contributed to the short life expectancy of our early ancestors. We would have started having babies in our early teens. By the time those fortunate enough to have reached 30 would have been considered the elders of the tribe.

Medicine did not exist, nor did dentistry. With no knowledge of sanitation, many women and infants must have died in childbirth. Our Paleolithic forebears may have compensated for this somewhat by developing and orally passing on knowledge regarding the effects of various beneficial herbs.

Even when Paleolithic hominids succeeded in finding ample food sources, they were faced with a formidable problem—locating shelter. Without sophisticated metal tools, building anything like a modern house would be out of the question. The most sophisticated human dwellings from the late Paleolithic were houses made from animal bones and antlers. Perhaps animal hides, which have not survived in the fossil record, were used to insulate these shelters as well as clothe our ancestors.

Earlier, human bands must have competed for the most suitable caves in mountainous regions. Not only would our ancestors have to contend with other human bands and our Neanderthal cousins for such choice habitats; they also would have to compete with such formidable cave-loving creatures as bears and huge felines. It is inevitable that humans were not always victorious in these contests.

Then there was the small matter of mobility and tribe size. Draft animals were far in the future; all land travel was by foot. If you couldn't get along with the few dozen people in your band, tough luck! The nearest neighboring tribe might be far beyond your walking range.

Let's Try the New Stone Age: The Neolithic Era

It is unlikely that most readers of this book would find Paleolithic life enticing. So let's skip forward a few millennia to the New Stone Age. This Neolithic Era extends from about 10,000 to 4,000 B.C. During this era, at least some people lived in mud-brick houses. Although these were not equipped with indoor plumbing and lighting, they at least kept the elements out. Neolithic food was a bit better and more dependable than Paleolithic. Agriculture was being developed, and some crops were farmed. Instead of chasing fleet game animals with spears and bows, our Neolithic ancestors could choose a tender morsel from the domesticated beasts and fowl.

However, sanitation was still primitive. The odors alone from dumped slops in Neolithic towns such as Catal Huyuk and Jericho must have been intense as local population levels exceeded a thousand.

Medicine and dentistry had not advanced. Lifespan had not substantially improved from Paleolithic levels. And there was a new wrinkle to add even more difficulty for the human participants: organized warfare had been invented. Probably, the first wars were not between rival agricultural settlements. Instead, the comparatively rich lifestyles of the agriculturists and animal herdsmen must have attracted the unwanted attention of nomadic Paleolithic tribes. Early on, Neolithic towns constructed defensive walls to keep the nomads at bay.

Human land mobility, though, had increased since some domesticated creatures—notably the ox—could be used as draft animals. During the late Neolithic, humans may have used the recently developed sail to venture out on the open seas. In the eastern Mediterranean, village culture began to spread from the mainland to off-shore islands.

Although Neolithic humans ate better and lived more securely than their Paleolithic cousins, life was still harsh and short, and farming and building used only stone tools. So let's jump forward a bit further in time from the Neolithic dawn.

The Bronze Age: Civilization is Born!

About 6,000 years ago, some unknown genius discovered that certain metals could be processed and altered with the aid of fire. With this bit of mastery over the physical world, numerous possibilities arose for our ancestors. Villages grew into cities. With the use of metal tools, hydraulic systems were developed to supply and store fresh water and to remove sewage. In some of the new cities, these advances led to populations of 10,000 or more.

Humans ate better, too, as metal tools allowed farmers to till the land and harvest crops with greater efficiency. The new technologies allowed for the construction of impressive, multi-storied buildings.

Writing was developed, which led to legal codes and centralized governments. In the new cities, various social classes arose. You would do well if you were born into the ruling caste, but not so well if you started your life as a slave.

With their newfound powers, responsible rulers strove to distribute agricultural produce fairly among the populace and to store food against the famines and droughts that would surely follow the abundant "seven good years." For the first time, some people had the leisure to think about the meaning of existence, the afterlife, and other related topics. Religious cults of Earth goddesses and sky gods gained followers in many parts of the world. Human sacrifice and ritual self-mutilation were not unknown.

In certain regions—notably the Aegean—a golden age of peace and prosperity occurred. But it should not be supposed that such an occurrence was due to improvements in human nature. In all likelihood, the naval technologies of the Cycladic Islands temporarily protected local populations from the depredations of marauding armies equipped with chariots drawn by the newly domesticated horse.

As population increased, civilization began to spread from its original birthplaces. In some regions, city-states joined to form the world's first empires. To provide hot water to increasing populations, many Bronze Age towns were constructed in tectonically active regions—not a very good idea from our perspective!

For the rich at least, medical care began to improve as written compilations of medical practice allowed for the training of new generations of practitioners. But the average life expectancy was still far below what we enjoy today.

The Bronze Age was a time of adventure and expansion. But for most at least, it was not a picnic.

The Age of Iron

About 3,000 years ago, people mastered the use of iron. Applied as weapons, this strong metal led to the growth of organized armies. Empires spread across the globe and strove against each other. In the west, Persia was supplanted by Athens, which itself was ultimately absorbed into Rome. Across the tortuous Silk Road, trade flourished between the Romans and the growing eastern empire of China.

A network of roads was constructed and cities that could house 100,000 or more. It seems unlikely that the innovations of the earlier hydraulic engineers of the Bronze Age Aegean could keep pace with the population growth. Perhaps because of population growth and poor sanitation, plagues were not uncommon. It was much more likely that a random person would die horribly from a plague than as a result of the incessant warfare.

The old gods and goddesses had failed. They could prevent plagues and wars no better than they could forestall earthquakes and volcanic eruptions. Faced with the choices of mindlessly worshipping these ancient deities or participating in the mind-drugging, bloody circuses supplied by the ruling elites, it is not surprising that many elected to become monotheists. The spread of religion and ethical philosophy in this period was greatly influenced by the adoption after 1,000 B. C. of alphabetic scripts, which greatly enhanced literacy. Art and literature flourished as well—at least for the affluent. It is difficult to imagine that the average person would have had time for such luxuries.

Ruins dated to the Iron Age abound in western Europe. In some places, buildings, highways, and aqueducts constructed in that period are still in use.

The Middle Ages: A Time of Turmoil

The cause might be the failure of the ancient gods or it might lie with the successful incursions of the barbarian hordes. Perhaps climate change played a role as crop failures made organized military adventures more difficult. Whatever the ultimate reasons, the organized bureaucratic structures of the unified Roman world began to decay and disappear before A.D. 500.

Population plummeted and literacy declined in Western Europe. Where peaceful farmers had gathered the fruits of the field, where philosophers and scholars had debated ethics and physics, scattered bands of survivors struggled against each other and the elements in a world without large-scale organization.

One of the surviving fruits of the late classical world was monotheism. If western civilization had not collapsed when it did, monotheism might have become a unifying force. Sadly, this was not to be the case. A widespread belief in one creator soon degenerated into a series of competing branches of monotheism, each maintaining the same sacred scripture, but each with its separate interpretation. Not only did Christians, Jews, and Moslems fight with each other; vast quantities of blood was also spilled in strife among rival sects of Christianity and Islam.

One thing that united the medieval monotheists, however, was a hatred of the polytheistic religions that had preceded them. Woe to the independent woman who helped preserve some ancient knowledge of herbal medicine—she was likely to be declared a witch and burned at the stake!

With the primitive sanitation and medical practice, plagues caused by human stupidity and superstition ravaged the population. But to be fair, there were some bright spots in the midst of this medieval murder and mayhem. Visual art had originated during the Paleolithic as a form of sympathetic magic. Cave paintings of mammoth and other large herbivores was then thought to increase the herds and the hunters' yield. Painting and sculpture developed further during the Bronze and Iron Ages, resulting in masterpieces found in the collections of the world's great museums. But in the Middle Ages, when illiteracy was widespread and the new religions required visual methods to teach scripture to the masses, painting and related arts had a significant flowering. Many of these works have survived and can be enjoyed on the walls of cathedrals, mosques, and synagogues constructed during that period.

The visual arts were also used to illuminate hand-copied classical manuscripts that had survived the fall of Rome. We owe a great deal to the scribes and manuscript illuminators of that era.

Music also advanced during the Middle Ages, perhaps for the same reasons as the visual arts. Choirs perform hymns composed by such notables as Hildegard of Bingen to this day.

If you were very lucky during the medieval era, you might have worked as an architect, artist, or composer. Or you might have led a sheltered life as a scribe in a protected religious enclosure. But for most people in that period, life was squalid, uncertain, and short.

The Renaissance: Civilization Awakens

Around A.D. 1400, the European Renaissance began, and it is widely viewed as a reawakening and revival. Perhaps because of the recently developed printing press, more books were published. Literacy and learning increased after languishing for a millennium.

The practice of modern science began. Humans learned that their world circled the Sun and was not the center of the universe. Scientific principles—anatomy, the camera obscura, and perspective—were applied by prolific painters and sculptors to their rapidly evolving disciplines.

In both the west and far east, the technology of the sailing ship reached new levels of perfection. Trade, exploration, and colonization became global enterprises.

However, much of this advance came with an increase of international turmoil. If the Ottaman Turkish conquest of the remains of the ancient Eastern Roman Empire had not precipitated a large-scale migration of eastern scholars to the Italian city states, the renaissance may not have happened. The schism of western Christianity into Protestantism and Roman Catholicism resulted in some of the bloodiest wars that humanity had known.

Was There Ever a Golden Age?

Our search through history for a golden age comes up empty. In every previous era, things might have been exciting and glorious for the lucky few. But for the majority in most past eras, life was more uncertain than it is today.

We should learn from the past. We should apply the wisdom of our ancestors and avoid their follies. Bur we should never forget that humans are the "movers and shakers of Earth." It is our obligation to look towards the future and apply our arts and sciences to ensure a good life for all the peoples of our emerging global civilization.

Any attempt to restore our planet's ecology that ignores the human element will fail. If the majority of people are poverty stricken or victimized, ecological awareness will fail. Since ours is the first era in which a substantial fraction of the human population lives well, ours is the first with a chance to restore the planet. Let's rise to the challenge!

Further Reading

Many authors have considered the laborious development of human culture and civilization, the challenges and capabilities of our ancestors. For a survey of prehistory and history through the Iron Age, see Jacquetta Hawkes, *The Atlas of Early Man* (New York: St. Martins Press, 1976). Another wonderful source is Jacob Brownowski, *The Ascent of Man* (Boston: Little, Brown, 1973). Finally, no student of civilization's development can ignore Kenneth Clark, *Civilization* (New York: Harper & Row, 1969).

Part III
Sky Harvest

Chapter 11
Raw Materials from Space

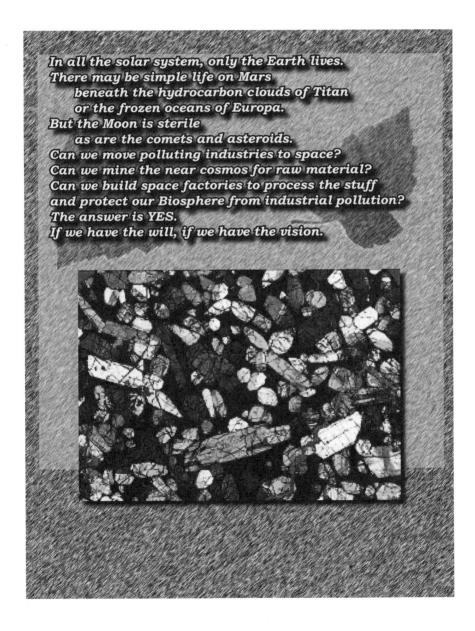

In all the solar system, only the Earth lives.
There may be simple life on Mars
*   beneath the hydrocarbon clouds of Titan*
*   or the frozen oceans of Europa.*
But the Moon is sterile
*   as are the comets and asteroids.*
Can we move polluting industries to space?
Can we mine the near cosmos for raw material?
Can we build space factories to process the stuff
and protect our Biosphere from industrial pollution?
The answer is YES.
If we have the will, if we have the vision.

G. Matloff et al., *Harvesting Space for a Greener Earth*,
DOI 10.1007/978-1-4614-9426-3_11, © Springer Science+Business Media New York 2014

> *"Allons! We must not stop here,*
> *However sweet these laid-up stores, however convenient this*
> *dwelling we cannot remain here,*
> *However sheltered this port and however calm these waters*
> *we must not anchor here,*
> *However welcome the hospitality that surrounds us we are*
> *permitted to receive it but a little while."*
> —From the poem "Song of the Open Road" by
> Walt Whitman

Walt Whitman was right: Earth is not a closed system. The road to the riches of the Solar System lies open. Nature has provided humanity with a virtually unlimited supply of raw materials, and that supply is right above our heads. The asteroids, comets, planets, and moons of the Solar System contain enough of the soon-to-be-scarce raw materials required to feed our technological civilization for thousands of years.

As the nations of Earth increase their consumption of raw materials to supply our expanding global civilization, prices will inevitably rise due to the economic laws governing supply and demand. Earth has a finite supply of fossil fuels, copper, zinc, titanium, and other elements as well as the minerals composed of them. In a peaceful world, competition for these resources will lead to higher prices, lower consumption, and recycling of existing supplies. If these approaches fail, as they often do, armed conflict may result. In an age where many have weapons of mass destruction, the consequences are unthinkable.

What are the resources of the Solar System? Where do we find them? How do we retrieve them? What do we wish to do with them? These questions are addressed in this chapter.

An Inventory of Solar Riches

In considering the riches of the Solar System, it is convenient to divide it into a number of discrete regions, not too dissimilar to the divisions of the living Earth into various discrete biospheres.

The Sun

First, we discuss the Sun, which may be considered as a "zone of fire." Spewing from the Sun into the interplanetary medium is the highly variable solar wind. Moving at 200–800 km/s, this stream of ionized hydrogen and helium nuclei and electrons has a typical (but highly variable) density of 10 particles per cubic centimeter, at Earth's solar orbit.

Space miners might consider using hydrogen from the solar wind to produce water (in combination with oxygen) for use by lunar colonists. A rare form of helium—helium-3—is also present in the solar wind and could theoretically be retrieved for use in terrestrial fusion reactors.

However, there are problems. To mine the solar wind on a practical scale, one requires an electromagnetic scoop about 1,000 km in diameter. Although the superconducting materials required to maintain the scoop function work well far from the Sun, today's superconductors will certainly fail in the inner Solar System, where temperatures and solar-wind densities are high. So barring breakthroughs in material science, direct mining of the solar wind remains an option for the far future.

Mercury and Venus

Moving out from the central solar fire, we next come to the realm of the inner planets Mercury and Venus. At one-third Earth's solar distance, Mercury is a challenging destination for contemporary spacecraft. Multiple trajectory-altering passes of Earth, the Moon, and Venus is one way to project a spacecraft on a Mercury-bound path. Once we've arrived, are there any useful Mercurian resources?

Surprisingly, the answer to this question seems to be "yes." Billions of years ago, water-bearing comets repeatedly crossed the inner Solar System, sometimes impacting the inner planets. Even though conditions on small, airless, and hot Mercury do not seem ideal for water retention, Earth-based radar studies reveal water-ice deposits near the planet's poles. These are probably comet-impact remnants in the interior of craters protected from sunlight. Such water may be of interest to an eventual Solar System civilization, but seems of little importance to water-rich Earth.

We'll skip torrid, greenhouse Venus in our resource inventory. Resources may certainly be present there, but retrieving them from this inhospitable world is problematical.

Earth's Moon

Humans have visited in person only one Solar System world beyond Earth. This is our planet's large Moon. At present, many spacefaring nations or transnational entities—the United States, Russia, Europe, Japan, China, and India—are enhancing their capabilities so that astronauts can eventually visit our Moon again. Certainly national pride is a factor here, and we can expect flags of many nations to be raised above the silent lunar surface. But it is possible that the new Moon efforts will also contribute something to terrestrial civilization.

Katharina Lodders and Bruce Fegley, Jr., tabulated estimates of lunar composition, largely based upon analysis of lunar samples returned by Apollo spacecraft. The Moon's mass is about 44 % oxygen, 19 % magnesium, 6 % aluminum, 20 % silicon, 7 % calcium, and 3 % iron, with trace amounts of other elements.

Since the Moon and Mercury are similar in many respects, it is hoped by many that deposits of water-ice will be confirmed in lunar craters near the poles, protected from direct sunlight. Until recently, evidence for such a resource was ambiguous at best.

Although trace amounts of water had been detected in samples returned from the Moon by Apollo astronauts and the Soviet robot Luna 24 during the late 1960s and 1970s, many researchers concluded that the likely cause was contamination. During the 1990s, inconclusive searches for lunar water were conducted using the U.S. military space probe Clementine, NASA's lunar prospector and the giant radio telescope at Arecibo, Puerto Rico. After 2000, additional inconclusive searches for lunar water were conducted using probes launched by China, Europe, and Japan.

The most promising results to date were obtained in 2009–2010 by research teams in India and the United States. In September of 2009, the spectral signature of water was detected in reflected sunlight from the lunar poles. In November 2009, after NASA's Lunar Reconnaissance Orbiter had competed its mission and was intentionally impacted near the lunar polar regions, water vapor was detected in the ejected plume from the impact site. In March 2010, it was announced that the Min1-RF experiment aboard Chandrayan-1 had discovered 40 dark craters near the lunar north pole that contain more than a trillion pounds of water-ice.

In 2017, NASA hopes to deploy a lunar robotic rover specially tailored to search for water. This is a preliminary step to locating the best site for a human lunar base from which the techniques of tapping lunar resources can be further developed.

As tabulated above, the Moon is approximately 20 % silicon by mass. By happy coincidence, silicon is the major element in certain varieties of photovoltaic cells. So it is not impossible that future astronauts could mine silicon from the lunar crust to construct giant arrays of photovoltaic cells (Fig. 11.1). This energy could be converted to microwaves and beamed back to Earth to supply a substantial fraction of our planet's energy needs.

Sun power from space is discussed in greater detail in Chap. 12. It is enough here to mention that large craters near the lunar limbs could be covered with thin-film silicon photovoltaic cells. If care was taken, the lunar face would not be greatly altered by this process.

Solar energy would first be converted into electricity and then into microwaves. These microwaves could either be transmitted directly to Earth receiving stations or to relay satellites near the Moon. Conceivably, a lunar base or colony could support itself by selling solar power to energy-hungry terrestrials.

Deep within the Sun, nuclear processes combine hydrogen atoms to form helium, releasing enormous energy in the process, providing the power that keeps the Sun shining and Earth warm. Some of this helium is in a rare form called helium-3. Helium-3 contains two protons and only one neutron rather than the usual two neutrons. While rare on Earth, helium-3 should be rather plentifully embedded

Fig. 11.1 Artist's conception of a solar-powered lunar base and mining facility (courtesy of NASA)

within the surface dirt, or regolith, on the Moon after its bombardment for billions of years by the solar wind.

Why is this important? Helium-3 might be the nuclear fuel we need to make mass-scale nuclear fusion reactors possible, providing relatively clean and safe nuclear power to future generations on Earth. As stated in Chap. 9, electrical power generated by breakeven nuclear fusion is years away from viability, and may always remain so unless something comes along to change the way we are approaching the technical problems. Helium-3-based fusion might be the disruptive technology we need to make fusion power viable.

It is easy to demonstrate that only a small amount of helium-3 would be enough to supply our civilization's entire energy requirement. Consider a scenario in which we require globally 30 trillion (3×10^{13}) watts of electric power. Furthermore, assume that it is all to be supplied by fusion reactors that burn a mixture of heavy hydrogen (deuterium) and helium-3. Advantages of this thermonuclear fuel mixture include relatively easy ignition and low waste radiation. Every year, about 10^{21} J of fusion energy (E) are required. If we next apply Einstein's famous mass-energy equation, $E = ?mc^2$ (where ? is a fraction of reacting mass m that is converted into energy in our fusion reactor and c is the speed of light) for the case of thermonuclear fusion, which converts a maximum of 0.4 % of reactant mass into useful energy, we find that only about 1 million kilograms per year of helium-3 is required by a reactor technology burning 50 % helium-3 by mass.

Current fusion reactor research is primarily concentrated on the combining of heavy hydrogen atoms to produce helium, releasing energy in the process. The energy would then be converted into electricity. The reactions currently in development release a significant amount of their energy in the form of radioactive neutrons. These neutrons represent a significant health risk and must be managed accordingly. In contrast, a fusion reaction involving helium-3 and deuterium would produce much less residual radioactivity. It is possible to envision hundreds of relatively clean fusion reactors providing electrical power for our cities with their fuel being imported from the Moon.

Current estimates place the amount of helium-3 on the Moon at about a million tons. This should be enough to meet the world's energy demands for thousands of years, even if we don't develop additional alternative clean energy sources. Unfortunately, obtaining the helium-3 from the Moon may not be easy. Since the atoms of this substance are embedded in the lunar regolith, the surface soil on the Moon will have to be stripmined and processed to release the helium. To accomplish this, an infrastructure will have to be erected there to allow the mining, transport, heating, and redistribution of tons of surface soil.

Unlike on Earth, where stripmining permanently mars the surface and destroys local ecologies, the impact on the Moon will be extremely minor. The Moon's surface is already dead, scoured of any potential life by the relentless solar radiation baking it for 14 days at a time. Gathering up the soil and redistributing it will damage no lunar ecologies, for there are none to damage. Using this resource from the sterile depths of space will help preserve ecologies here on the living Earth.

There is one catch to lunar mining and settlement plans. (In space, there is almost always a catch or two!). An omnipresent layer of dust bothered Apollo astronauts trudging or riding across the lunar surface. Remote videos of Apollo lunar module liftoff from the lunar surface observed that rocket blasts blew up so much dust that the flags planted by astronauts actually waved in the lunar vacuum. This material coated the astronauts' spacesuits. The interior of the lunar modules became dust repositories. Future lunar colonists and miners will find this dust layer to be a source of inconvenience and frustration.

Compounding the problem is the low lunar surface gravity. Large-scale lunar mining may raise lots of dust, and because of the low gravity, lunar dust will not settle rapidly. Not only would uncontrolled lunar mining raise a dust layer inconvenient to terrestrial lovers of the Moon, it might negate plans of astronomers to operate observatories on the lunar farside. Clearly, lunar development must be carefully thought out. But beyond the Moon and well within reach of planned interplanetary spacecraft is another source of extraterrestrial resources.

Near-Earth Objects: If We Gotta Move Them, Why Not Use Them?

Between Venus and Mars, there is a class of rocky, stony, and icy objects that orbit the Sun. Some are the size of boulders, others are as large as mountains (Fig. 11.2). These Near Earth Objects (or NEOs) are either asteroids or extinct comets. As

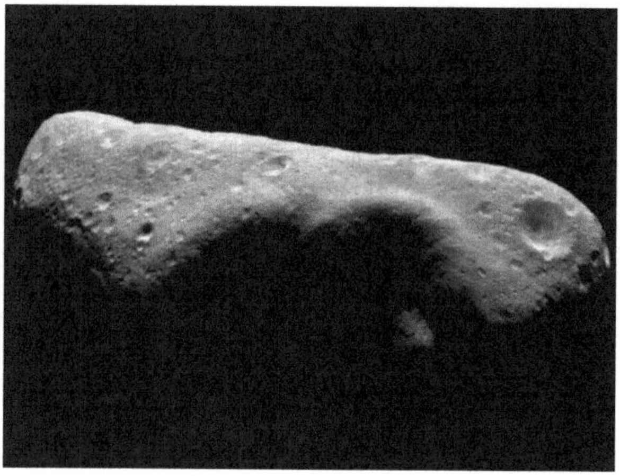

Fig. 11.2 Spacecraft image
of Eros, a large NEO
(courtesy of NASA)

discussed in Chap. 14, they occasionally whack Earth with devastating conse-
quences. To protect our planet and ourselves we must track and categorize these
NEOs. Then we must investigate and develop techniques to divert the trajectories of
Earth-threatening NEOs. But if we have to move them, why not use them?

Spacecraft have visited asteroids before, and from each such mission, we have
learned a great deal about their composition. The NASA Galileo spacecraft imaged
asteroid 951 Gaspara in 1991 and asteroid 243 Ida 3 years later. In 1997, the NASA
Near-Earth Asteroid Rendezvous (NEAR) mission captured images of 253 Mathilde
and visited asteroid Eros in 1999, ending its mission by gently crashing into Eros'
surface. The Japanese explored asteroid 25143 Itokawa with their Hayabusa spacecraft in 2005 and returned a small sample of material from this object to Earth.

According to the Roskill Information Services, world production of iron
exceeded 1.4 billion tons in 2006. The rapidly expanding economies of China,
India, and Russia accounted for much of this total. The demand for iron ore is
expected to increase each year into the foreseeable future, with more mining and
recycling required to meet it. A relatively small asteroid approximately 1 km in
diameter might contain 2 billion tons of iron ore—enough to meet the annual global
demand for 1 year. The asteroid 16 Psyche may contain enough iron ore to meet
global demand for millions of years. And that is just the iron ore. Spectroscopic
analysis of other asteroids shows that they contain nickel, cobalt, copper, platinum,
and gold.

Although robots from Earth have flown by or touched down upon a few NEOs,
one of the best tools for remotely observing them is the huge radio telescope located
in a crater in Arecibo, Puerto Rico. To observe certain characteristics of an NEO, a
radar pulse is sent out from this instrument. The reflected return pulse is later
received and analyzed.

We now know that there are a few thousand of these objects in the 100-m size
range or larger. They rotate with periods of hours or days. Some NEOs (and main-
belt asteroids) are known to have satellites.

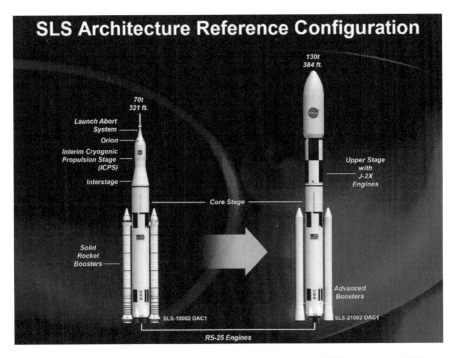

Fig. 11.3 Two versions of the SLS. The one on the *left* carries an MPCV (courtesy of NASA)

There seems to be a large variation in physical characteristics. Some NEOs are very dense. These are most likely composed of heavy metallic elements such as iron and nickel and are similar in composition to Earth's core. Others are stony, of lower density; in composition at least, these may not be dissimilar to Earth's crust. Some NEOs are very tenuous. These may resemble flying rubble heaps, held loosely together by the aggregate's weak gravitational field. Many NEOs are classified as chrondites. With water, carbon, and various volatile substances, these are most likely extinct comet nuclei.

Now that we have established that NEOs contain essentially limitless raw materials that our civilization so desperately needs, what do we do about it? How do we get access? Most importantly, how do we get the raw materials we need from there to here—affordably? Approaching the problem from an American point of view (since the authors are Americans), we can begin building our in-space mining infrastructure using the next generation of rockets and spacecraft under development by NASA and private industry to support a deep-space capability in the late 2010s and early 2020s.

The current NASA approach to voyages above low Earth orbit (LEO) is centered upon the Space Launch System (SLS) and Multi-Purpose Crewed Vehicle (MPCV). The SLS (Fig. 11.3) is scheduled to become operational in 2016–2017. It will come in several configurations and be capable of injecting 70–130 tons of payload to LEO. Although its primary function will exploration of cis-lunar and near-

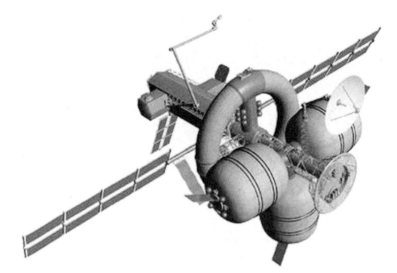

Fig. 11.4 Nautilus-D interplanetary spacecraft concept, showing centrifuge and solar panels (courtesy of NASA)

interplanetary space, it could also be used to re-supply the International Space Station.

The MCPV, which is also undergoing development, will carry a crew of 2–4, have an EVA (extra-vehicular activity) capability, and the ability to support a crew for 21 days. Like the Apollo spacecraft of the 1960s and 1970s, the MPCV is configured for a water landing. Its total on-orbit mass will be about 21,000 kg.

For flights to the vicinity of the Moon, the 21-day mission time limit of the MPCV will be ample. But for more extended voyages to NEOs, the crew will require more interior space and provisions. A true interplanetary craft that could be used for voyages to NEOs and beyond is the NASA Nautilus-D concept (Fig. 11.4). This solar-powered craft could be lofted by several SLS launches and would include such amenities as a centrifuge to allow the crew to counteract the effects of micro-gravity during their long duration (6 months to a year) round-trips to nearby NEOs.

Since the current political/economic climate favors commercial space activities, a number of private space enterprises have come up with competing concepts. The Space Explorations Technologies Corporation (Space-X) has to date (November 2012) completed two successful unmanned flights to the International Space Station using its Dragon capsule launched using a Falcon-9 rocket. Versions of the forthcoming manned version of the Dragon could be outfitted for NEO visits. The interplanetary craft could be constructed in orbit using several launches of a heavy-lift version of Falcon-9 called Falcon-9 heavy. To reduce on-orbit mass (and cost), the interplanetary craft could consist in part of inflatable modules such as those already tested in space by Bigelow Aerospace and planned for an eventual hotel in LEO.

Another entry in the U.S. NEO Olympics is Excalibur Almaz, Inc. This private space company, which is based on the Isle of Man with offices in Houston and Moscow, is partially funded by NASA. At the end of the Cold War, Excalibur succeeded in obtaining Soviet space-station hardware and reentry vehicles. With a forthcoming first launch using Russian Proton rockets, Excalibur intends to ultimately send tourists on circum-lunar trajectories and demonstrate the capability to send crews to explore near NEOs.

One scenario for NEO mining is to limit our NEO sample to those objects that might someday impact Earth. Radar-ranging and robotic space probes could then be employed to determine many physical properties of this NEO class.

Planetary Resources, Inc., a private Seattle-based initiative seeking to mine asteroids for profit, proposes a multi-pronged robotic approach. First, they will establish in LEO a space telescope dedicated to the NEO search and characterization efforts. This will be followed by a series of low-cost NEO-interceptor probes to fly by selected NEOs and engage in preliminary exploration. In the prelude to NEO mining, a series of robotic NEO-rendezvous prospectors will be dispatched.

After robotic NEO exploration commences, human crews might be dispatched to perform more elaborate investigations of the most promising NEOs. To reduce astronaut exposure to potentially harmful galactic cosmic radiation, only very near NEOs should be visited. Round-trip flight times (including stay times) of a few hundred days or less are reasonable.

Although the crews might be flown to the subject NEOs as quickly as possible, surveying and mining equipment could be delivered along more energy efficient, but more time-consuming, trajectories. After all, such equipment will have few worries about cosmic ray exposure.

After they reach their NEO destination, astronaut crews will rendezvous with the celestial object. Rather than landing on a planetary body such as the Moon, as was done by the Apollo lunar landing modules, astronauts will essentially dock their spacecraft with the low-gravity NEO. They would next establish a base camp, perhaps within a crater on the NEO. They would then utilize remotely operated rovers to conduct surveys of NEO surface and subsurface composition, seismic properties, etc.

The next order of business is Earth protection. Equipment, possibly resembling that discussed in Chap. 14, would be erected to slightly alter the NEO's solar orbit so as to reduce the threat to Earth.

Different strategies are required for different NEO classes. Altering the trajectory of a rubble pile requires different equipment than moving a metallic NEO in the same fashion. A number of devices exist for NEO trajectory alteration. The solar collector, discussed in Chap. 14, works best with extinct comet nuclei or NEOs with extensive dust layers. The mass driver—a type of electromagnetic catapult that ejects NEO material at high speed—would function well with rocky or stony NEOs.

As well as altering NEO trajectory, several additional dynamical options must be addressed by would-be space miners. Do we de-spin the NEO or reduce its rotation rate? Do we steer it into high Earth orbit where astronaut teams could deconstruct it in the vicinity of Earth? Or do we instead send portions of the NEO towards Earth, controlled by solar sails, solar-electric engines, or similar devices?

If we can steer an NEO into an Earth orbit with a perigee lower than about 2,000 km, a long and thin conducting wire can be deployed to interact with Earth's magnetosphere and produce additional thrust. A detailed explanation of the operation of such an electrodynamic tether can be found in the authors' previous book *Living Off the Land in Space* (Springer, 2007). Such a tether could be used to alter the NEO's Earth orbit with great precision, allowing for a controlled impact in an unpopulated but accessible region of our planet.

Finally, we might wonder what NEO resources are most profitably mined for use in space. Rock could be used as cosmic ray shielding by interplanetary spacecraft; water-ice could be used in the ecospheres of space habitats or electrolyzed as rocket fuel.

It seems most unlikely that the old science fiction scenario of space prospectors becoming rich after their discovery of a gold- or uranium-rich NEO will ever come to pass, even though spokespersons for Planetary Resources, Inc., hope to make money from platinum-rich NEOs. Simply delivering such an object to Earth (hopefully along a controlled trajectory) will essentially flood the market and greatly reduce the value of the precious, previously rare, commodity.

Earth's economy and climate might benefit if NEOs are mined for silicon and other materials from which solar power satellites could be manufactured. Use of such free-flying satellites to beam solar power to Earth has certain advantages over Moon-based power scenarios. NEOs could also be disassembled and reconfigured into sunscreens to reduce global warming, as discussed later in Chap. 16.

Comets

French Queen Matlida, in or around the year 1077, allegedly arranged the creation of the Bayeux Tapestry (Fig. 11.5). Seeking to honor William the Conqueror and the 1066 Norman invasion of England, Matilda commissioned the tapestry for placement in the Bayeux Cathedral. This 50-cm by 70-m work of art is rich in historical content, telling the story of the invasion and the events leading up to it.

In reality, and as chronicled on the tapestry, Halley's Comet appeared just before the invasion, signaling doom for the English. Surely this wispy celestial figure was an omen from God—at least that is the way medieval people interpreted comets and their often majestic appearances in the sky. In our day, we know far more about comets, where they come from and why they appear as they do in the skies above. Perhaps we too, however, should take a cue from the Bayeux tapestry and take their appearance as an omen—though not of doom, but of plenty.

Like asteroids, comets are often remnants of the early Solar System, orbiting the Sun for billions of years and often having no direct interaction with other Solar System bodies. Unlike asteroids, they are made from an ice-dust-rock ensemble that

Fig. 11.5 A man-at-arms warning the English ruler Harold of the disastrous omen of Halley's Comet. This image is a detail of the Bayeaux tapestry (ca. 1070–1080), Bayeux, France

produces a visible atmosphere when exposed to solar radiation. Comets are often spectacular and have been objects of study since the first humans observed them. Figure 11.6 shows comet Neat providing a truly awesome sight in the night sky.

The tail of a comet may extend millions of kilometers into space, always trailing away from the Sun due to the pressure of sunlight. By contrast, the nucleus itself is typically only tens of kilometers in diameter. That such a small object can produce a massive and beautiful display is truly one of nature's wonders.

Comets are abundant in the outer Solar System, and only a few of them have orbits that regularly bring them close enough to the Sun for us to observe directly. Most are quietly careening through the outer Solar System in nearly circular orbits, well beyond the orbit of Pluto. Some are in highly elliptical orbits (meaning that their orbits are shaped like eggs, rather than circles) that allow them to periodically plunge into and then out of the inner Solar System, providing us with the sky display described above. The most famous of these is Halley's Comet, which is set to make another appearance in 2061.

What resources will comets ultimately be able to provide for Earth? Perhaps none—at least directly. But any large-scale industrialization of space will require water. In fact, water may likely be the most important extraterrestrial resource required to support our expansion into the Solar System. Water is needed to support human life (drinking, food preparation, bathing) and our industrial activities (coolant, cleanser, solvent, and raw material). We also need it for fuel in our spaceships.

Fig. 11.6 Comet Neat shows its splendor in the Arizona sky (courtesy of NASA)

For instance, the main engines of NASA's now-retired space shuttle burned a combination of liquid oxygen and hydrogen, producing water as a by-product. That big pillar of "smoke" that was seen during shuttle launches was not smoke at all; it was a pillar of steam. Future spacecraft can be refueled using hydrogen and oxygen obtained from the electrolysis of water, perhaps made from cometary ice. (Electrolysis is the process in which an electrical current is passed through water, breaking it into hydrogen and oxygen.)

To provide water for use on the Moon, we might someday alter the orbit of a comet so that it smashes onto the Moon. The resulting ice fragments could then supply a human settlement there for decades or centuries. In a sense, we would be doing on the Moon what nature did to Earth so long ago. Earth's abundant water is thought to have arrived here with the impacts of comets into the surface billions of years ago.

Mars' Moons and Main-Belt Asteroids

Further out in this Solar System realm of rock and dust we come to Mars' tiny satellites Deimos and Phobos (Fig. 11.7) and the main-belt asteroids. These small bodies can certainly serve the ultimate needs of a space-based Solar System civilization.

Fig. 11.7 Mars satellites Phobos (*top*) and Deimos (*bottom*), imaged by Mars Reconnaissance Orbiter in 2007 (courtesy NASA)

However, there seems to be no immediate utilization of the immense resources in these objects to help reclaim Earth. So our best strategy might be to study and inventory the resources of these objects for the far future.

The same must also be said for Mars. If we elect to establish colonies on this barely habitable world and gradually transform its environment, Mars will be a net importer of resources for millennia.

Perhaps social and political adaptations of eventual Mars colonists could have a positive effect on terrestrial social and political structures. Such a scenario is not dissimilar from the effects of novel New World social structures (including participatory democracy) upon eighteenth-century Europe.

Realm of the Giants

Beyond the main-belt asteroids, at five times the distance of Earth from the Sun, is giant Jupiter. At 318 times Earth's mass, Jupiter is indeed king of the Solar System. However, the mass of our Sun is about 1,000 times the mass of this giant world.

Far from the Sun, we are now in the zone of gas and ice. Jupiter and its smaller colleagues, Saturn, Uranus, and Neptune, are giant gas balls. Jupiter and Saturn are mostly hydrogen and helium. Uranus and Neptune are rich in methane and ammonia. All of these planets have ring systems and are accompanied by many satellites. Many of the satellites of these giant worlds are coated with layers of water ice.

Some of these satellites are as large as small planets; others are captured asteroids or comets.

Space miners seeking to tap the resources of the giant worlds must also beware of multiple hazards. For example, Jupiter is equipped with a strong magnetic field and extensive radiation belts.

It seems unlikely that future space miners will venture this far from the Sun in search of water-ice or silicon for solar photovoltaic cells. But there is one tantalizing resource in the atmospheres of the giants that may someday be tapped for terrestrial use.

If we establish a nuclear-fusion based economy, a source of helium-3 would a good thing. And the atmospheres of the giants contain this isotope in approximate solar abundance—concentrations of about 1 part in 10,000.

In his book *Mining the Sky* (Reading, MA: Addison-Wesley, 1996), John Lewis describes conceptual plans to mine the upper layers of the giant worlds for this resource. A system of robotic shuttles would fly the multi-year round-trips between Earth and the giant world of interest. Suspended by balloons, helium mines would be dropped into the upper atmospheric layers of the large worlds. On their return to Earth, the shuttles would be loaded with this precious form of helium. In light of the interplanetary capabilities currently under development by many nations, this load is far from enormous. However, only one Apollo-massed interplanetary shuttle loaded with helium-3 needs to reach Earth each month to supply global energy needs in this manner.

Choices

As we look further into the Solar System, we come to the enormous resource pools of the Kuiper Belt dwarf planets and the Oort Cloud comets. But most of this material is best thought of as a resource to help maintain a future Solar System civilization, rather than for import to Earth.

Our use of space resources will depend upon what choices we collectively make as citizens of a global civilization. If we desire a solar-powered world, two choices are lunar and NEO resources. Do we develop the Moon because it is close and we have visited it in the past? Or do we instead bypass the Moon because NEOs must be moved around to reduce the risk of collisions with Earth—and if we have to move them, why not use them?

Alternatively, a substantial fraction of our planet's future energy requirements may be produced by thermonuclear fusion. In such a case, if breakthroughs do not allow tapping the solar wind directly and if lunar concentrations of helium-3 are too low, robotic exploitation of the giant planets may become a reality.

Unfortunately, we have no idea what direction humanity will take. But it's comforting to know that the resources we need are available in space and there are many options to obtain them for Earth's benefit.

Further Reading

For more about the possibility of future wars over natural resources, we recommend *Resource Wars: A New Landscape of Global Conflict*, by Michael T. Klare (New York: Holt, 2002). A good source of data on Solar System objects is Katharina Lodders and Bruce Fegley, Jr., *The Planetary Scientist's Companion* (New York: Oxford University Press, 1998). One of the best sources on Solar System resources is the very readable John S. Lewis's *Mining the Sky* (Reading, MA: Addison-Wesley, 1996). A somewhat more venerable treatment of the same topic, authored by an ex-astronaut, is Brian O'Leary's *The Fertile Stars* (New York: Everitt House, 1981).

Technical treatments of NEO resources and NEO mining possibilities can be found in two scientific papers published in "Near-Earth Resources," (Gertsch, Remo, and Sour Gertsch) and Mining Near-Earth Resources" (Gertsch, Sour Gertsch, and Remo), published in the proceedings of the U. N.-sponsored conference at which they were presented: John L. Remo, ed., *Near Earth Objects, Annals of the New York Academy of Sciences*, 1997, Vol. 822.

Chapter 12
Power from the Sun

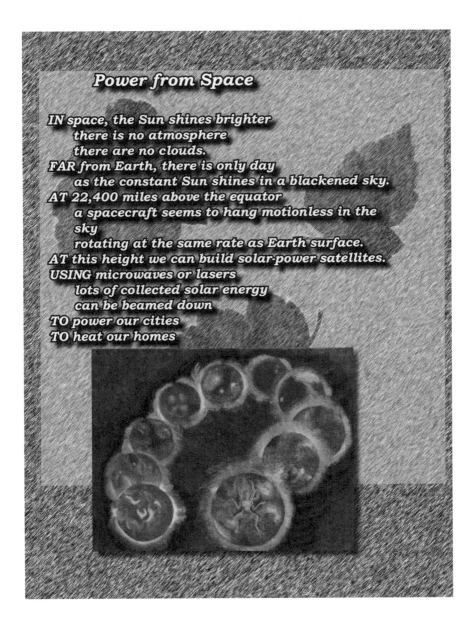

G. Matloff et al., *Harvesting Space for a Greener Earth*,
DOI 10.1007/978-1-4614-9426-3_12, © Springer Science+Business Media New York 2014

"Busy old fool, unruly Sun,
Why doest thou thus,
Through Windows and through curtains, call on us?"
—From the poem "The Sun Rising" by John Dunne

386,000,000,000,000,000,000,000,000 W. By any measure, that is a lot of power, and that is the Sun's power output every second of every day (Fig. 12.1). To make this number easier to work with, scientists write it as 3.86×10^{26} W, or 386 followed by 24 zeros. The amount of solar energy per second reaching Earth, which is 93 million miles from the Sun, is 1.74×10^{17} watts, or approximately 1,368 W/m^2. By way of comparison, in 2005 the total power output of the human race was approximately 1.5×10^{13} W! In that year, we generated a mere 0.009 % of what the Sun sends to Earth each second. If we can tap this tremendous energy source, then surely the global energy problem can be solved.

The truth is, we do use this energy already. All living things on Earth derive their energy from the Sun. though we humans do so indirectly. In photosynthesis, Earth's plants convert sunlight to carbohydrates for use in their growth and for their ultimate consumption as food by animals. We are near the top of the food chain, yet all of the foods we eat derived their stored energy from the Sun. Do you like steak? The cow from which the steak is cut consumed some sort of grain, which grew using sunlight in the photosynthesis process.

We use the Sun's energy in other ways as well. Fossil fuels are formed by the long-term decay of plants and animals that existed on Earth in its ancient past. The energy they collected during their lifetimes is stored in the coal, oil, or natural gas—all of which are called fossil fuels, since they are derived from fossils. When we burn the oil, we are releasing the energy stored chemically in that ancient plant. Think of it as a battery.

The recent trend toward biofuels is yet another example of how we use the Sun's energy for power. Biofuels are simply artificial fossil fuels that are manufactured from plant mass. The plants are grown, harvested, and processed to extract fuel that we then burn to run our machines. Unfortunately, the process of manufacturing biofuels involves many different steps, all of which are relatively inefficient. The end product, which is the output of whatever the machine burning the biofuel is intended to accomplish, is produced at the expense of a relatively large energy input.

According to the U.S. Bureau of Land Management, the average American household uses approximately 14,000 kilowatt-hours (kWh) per year. A kilowatt-hour is the amount of electricity a 1-kW appliance would use if left running for 1 h. For example, a 100-W light bulb burning for 1 h would use 0.1 kWh; if it were left on for 10 h, it would use 1 kWh.

Although this is currently far more power per person than that consumed by the citizens of the rest of the world, there is reason to believe that the rest of the world's

Fig. 12.1 The Sun emits an enormous amount of energy, enough to meet our energy needs for as long as the Sun and Earth exist (courtesy of NASA SOHO project)

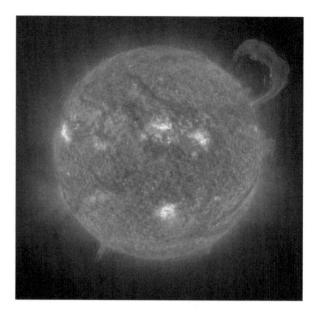

per capita energy consumption is rising and will continue to rise in the future as standards of living increase. Burning fossil fuels in automobiles, electrical power plants, or other machines such as jet aircraft currently consumes most of this energy.

We can convert sunlight directly into electrical power using a solar cell. Solar cells use quantum mechanical effects to convert some fraction of the Sun's energy falling upon them to usable electrical power. Terrestrial solar cells can be used effectively for applications that do not require a lot power, such as highway road signs and calculators. If you need megawatts on a continual basis, they are simply impractical in many locations. Terrestrial solar cells can generate power only during the day, only when the Sun is not obscured by clouds and rain, and in locations that don't degrade the materials from which they were made. Unfortunately, there is no place on Earth that gets continuous sunlight and never has bad weather.

It is well worth considering the ultimate potential of terrestrial solar power. After all, this is a very rapidly evolving technology that can be applied to satisfy either centralized or individual household energy requirements. Solar energy facilities are generally solar thermal (in which focused and concentrated sunlight heats a circulating fluid) or solar photovoltaic (in which sunlight is directly converted into electricity). Although the efficiency of solar energy conversion in existing facilities is generally 20 % or lower, future technological advances may raise conversion efficiency to 40 % or higher.

In Earth orbit, a flat plate facing the Sun is impacted by about 1,400 W/m^2 of sunlight. If you factor in the day/night cycle, the fact that the Sun is rarely directly overhead (even at the equator), and clouds often block sunlight, it is reasonable to assume that a typical solar facility on Earth's surface will receive an average solar power of about 100 W/m^2. If a future global power facility desired to satisfy a substantial fraction of human civilization's power requirements (say, 10^{13} W or 10 million megawatts) using centralized solar power at 40 % efficiency, the required collector surface area will be about 25,000 km^2. This corresponds to a square with dimensions 500 × 500 km (or 300 × 300 miles). Although this represents a small fraction of terrestrial land area, it must be acknowledged that residential and agricultural requirements in heavily populated nations are acreage scarce. Also, the best sites for locating huge terrestrial solar array farms are often far from population centers, which would necessitate extensive alteration to local and global power distribution networks.

What if you could locate these massive solar array farms in a location that is in almost perpetual sunlight with no cloudy or rainy days? You might have a chance at providing continuous power to an energy-hungry population without directly generating pollutants or emitting greenhouse gases. Then you might have a power solution worth considering. Space provides an optimal location to generate electricity for a power-hungry Earth.

In 1973, Peter Glaser proposed using microwaves to beam such space-generated power back to Earth. Others subsequently proposed using lasers or other electromagnetic radiation of various frequencies to beam the energy to Earth for conversion and distribution to the electrical power grid.

What are microwaves? Like sunlight and radio waves, microwaves are simply a part of the electromagnetic spectrum. The only difference between a microwave and a radio wave (from your favorite radio station) is its frequency. In fact, a microwave, a radio wave, and the light emitted by a reading lamp are all essentially the same thing, distinguished only by their different frequencies. All are considered electromagnetic radiation, and, as evidenced by the myriad radio transmissions passing through our bodies all the time; many are quite harmless to biological organisms such as humans and most other denizens of the biosphere. And the microwaves used by space solar power satellites can be designed to be equally benign.

Before going into the details of how these systems can be constructed and operated, the overall concept needs to be described.

High above the surface of Earth, very large panels of solar collectors can be deployed to convert sunlight into electricity. There are orbits around Earth, described later, which are in nearly perpetual sunlight. In these orbits, spacecraft can continuously generate electrical power without having to worry about the weather or that pesky darkness that happens every night. The power they generate is then converted to microwaves and beamed back toward Earth. After traveling through the atmosphere, these microwaves are collected by large antennas and converted back into electricity.

Figure 12.2 illustrates what a space-based solar power satellite might look like once it is operational. Generating power in this manner is the ultimate "green" energy. It consumes few resources from Earth, other than those used in its construction. The bulk of the system operates well out of the biosphere, and it does not blight the landscape as do conventional fossil-fueled power plants. The energy passes harmlessly through the atmosphere to be converted into relatively clean electrical power, which can then be used to power our cities and future-generation

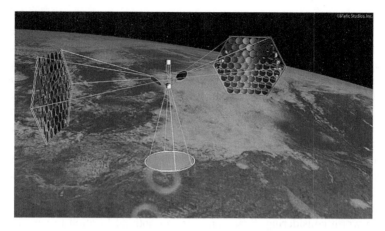

Fig. 12.2 Artist concept for an Earth orbiting solar power satellite beaming power back to Earth (artwork courtesy of the National Space Society)

electrically powered transportation systems—including our cars and trucks. The power source is the Sun, and it never sets for our orbiting power stations. Unfortunately, there are still some significant technical hurdles to overcome before this vision can become a reality.

To understand these hurdles, and some options for addressing them, we'll take a closer look at each part of a space-based solar power system.

The Orbit

To achieve the goal of nearly perpetual sunlight, a spacecraft cannot be in low Earth orbit. For reference, the International Space Station is located in low Earth orbit at an altitude of about 400 km. It is at about this altitude that all American space shuttle flights were conducted. Spacecraft in low Earth orbit circle the globe approximately every 90 min, experiencing a complete day/night cycle with each orbit as they do so. Spacecraft at these altitudes pass through Earth's shadow approximately 50 % of the time, and therefore they cannot produce continuous electrical power. There is an altitude, however, at which a spacecraft can orbit Earth and almost never enter its shadow. This particular orbit is called "geostationary" and it is located 22,300 miles above our heads.

A geostationary orbit provides to a spacecraft what its name implies. "Geo," meaning Earth and "stationary," meaning that it appears not to move as it circles the globe. This seemingly contradictory orbital situation results from the fact that Earth rotates, producing our 24-h day. In a geostationary orbit, the velocity required for a spacecraft is such that it circles Earth once every 24 h , matching the orbital rotation rate of Earth. The result is a spacecraft that always remains directly overhead at any particular point on Earth's equator. It is sufficiently far from Earth that it almost

Fig. 12.3 The International Space Station with solar array wings deployed (photo courtesy of NASA)

never enters Earth's shadow; thus there is no real "nighttime." A spacecraft in geostationary orbit will experience almost continuous daylight.

Imagine a 1-km wide band encircling Earth's equator in geostationary orbit. The sunlight falling onto this area of space, if collected for just 1 year, equals approximately the total amount of energy contained in all the remaining known oil reserves in the world today.

Methods for Converting Sunlight into Electricity

There are two good candidates for converting sunlight into electrical power. The first is the likely to be familiar to anyone who owns a solar-powered calculator or other electronic device. In this approach a semiconductor material in an array converts sunlight into electricity directly using quantum mechanical effects. The semiconductor is called a photovoltaic array (PVA), commonly referred to as a "solar array." The leading alternate approach is called solar dynamic power conversion. Solar dynamic systems use the energy in the sunlight to heat a working fluid that then drives some sort of electrical power generator. This approach is very similar to that used in coal-, oil-, and nuclear-fueled power plants on Earth.

Most spacecraft use photovoltaic arrays, converting sunlight into electrical power, and explore the inner Solar System using the power they generate. Over time, the efficiency at which these arrays operate has increased, allowing for either smaller solar panels or instruments that require higher power to be used.

The International Space Station (ISS), shown in Fig. 12.3, uses multiple solar array wings to produce its own electrical power. Each wing is 34 m (112 ft) long and 12 m (39 ft) wide. Now complete, the ISS has eight such wings that generate tens of kilowatts of power.

Space-based power conversion efficiencies are likely to be greater than 33 % by the middle of the second decade of the twenty-first century, making the technology very attractive for space solar power applications.

Sending the Power Back Home

Transmitting power through the air without wires is not science fiction. In 1964, pioneer William C. Brown demonstrated on live television that microwaves could power the flight of a miniature helicopter. In 2003, engineers at the NASA Marshall Space Flight Center flew a small, propeller-driven model airplane powered by a laser (instead of microwaves) sitting across the room. And in 2008, newspapers reported that Japanese researchers began testing a microwave power beaming system as part of their ambitious plans for eventually building solar power stations in space.

To make it work in our energy grid, the power generated by the solar arrays onboard the spacecraft will have to be converted to electromagnetic radiation and transmitted to a receiving station on the ground. In order to accomplish this, the frequency of the radiation transmitting the power must not be harmful to life, must not damage the atmosphere as it passes through, and must not be absorbed on its way to the surface by water, carbon dioxide, or other gases in the atmosphere. In addition, to prevent having extremely large antennas to broadcast and receive the power beam, the wavelengths of the radiation must also be as small as possible.

When all of these constraints are considered, the leading candidates for power beaming from geostationary orbit to the ground are microwaves. Microwaves can pass relatively easily through our water-laden atmosphere, they can be of a wavelength that is not absorbed by the human body, and they can be generated and received by antennae that are of sizes that can at least be mentally conceived.

A laser optimized for power beaming would have greater difficulty with attenuation as it traverses the atmosphere, but it would require less of an infrastructure on the ground. For example, the ground-based receiver would only be a few tens of meters in diameter versus a microwave rectenna with diameter of 1–10 km. If a method to decrease atmospheric losses is found and implemented, then laser power transmission would become the technology of choice.

Antennae for Transmitting and Receiving

As described above, space-based solar power stations will be sending the power back to Earth using microwaves. Recall that a microwave beam differs from a radio wave only by its frequency. They are fundamentally the same, except for their rate of oscillation (frequency) and wavelength. Radios need antennae to transmit and separate antennae to receive. In this case, the space-based transmitting antenna will need to be 1–2 km in diameter and the receiving antennae on the ground about ten times larger.

A note about building large structures in space is warranted. It is far easier to build large, even massive structures in space than it is to build similar structures on the ground. In space we are free from the unrelenting tug of Earth's gravity and can make our structures out of extremely low-mass materials that would not hold up under their own weight if they were used on the ground. Without wind and rain, we don't have to worry about many of the same size-limiting design constraints that we do in terrestrial applications. Prototype large structures have already been built in space.

The ISS became operational in the year 2000, and, now completed, it is approximately the same size as a football field. In the lobby of a building at NASA's Marshall Space Flight Center in Huntsville, Alabama, is a scale model of a football field (The University of Alabama versus Auburn University, of course), within which sits a same-scale model of the ISS. This allows casual visitors to know the scope of this grand engineering endeavor.

In addition to the space station, we humans have built and flown structures 20 km in length. In 1996, the Small Expendable Deployer System (SEDS) unfurled a 20-km-long, 0.075-cm diameter cable from a Delta II rocket. This feat was duplicated several times by NASA, and comparable feats were demonstrated by other spacefaring nations. This is certainly not a 20-km square solar array farm, but it is an essential first step toward the capability of building one in the future.

Distributing the Power

By comparison, this part of the system is simple. The power collected by the ground-based antennae must be regulated (voltage, frequency, and overall "quality") for connection to the power grid. Highly efficient power conversion and regulation technologies already exist and are in use throughout the world. Adapting them for use in space should not be difficult.

The Catch

Yes, there is always a catch. This virtually limitless, completely renewable, continuous power system will be terribly expensive to develop and launch into space. The launch requirements alone, at today's prices, are astronomical (pun intended). A power station large enough to supply the energy needs of a large city might weigh millions of kilograms and require many launches of our most capable rockets just to get the construction materials into space. Then it would have to be assembled—probably requiring humans since no matter how capable our robots may be, there is no substitute for having a person at the site in case something goes wrong.

At today's prices, the launch costs alone could be many billions of dollars (though with trillions of dollars used in 2009 to bail out the world's financial institutions, this seems like a worthwhile and affordable investment).

And then there are the rest of the infrastructure costs. How much will it cost to build that spacecraft and solar arrays, the antennae, and the ground support equipment? These costs could easily total in the billions of dollars. For lack of detailed accounting analysis, let's say this infrastructure cost is on the order of half the launch cost, placing the total system price at approximately $30 billion. For comparison, using today's dollars, a coal-fired power plant would cost "only" hundreds of millions of dollars.

Space-based solar power is clearly not cost competitive—yet. Improvements in solar array efficiency seem to be occurring on a regular basis. As of this writing, some inventors are claiming efficiencies greater than 35 %. Solar cell arrays are also becoming thinner and less massive—recent advances point to a possible operational array thickness of 10 μm or less. Previous studies assumed much lower efficiencies and much thicker arrays, which implied the requirement for more mass in orbit to supply a given power to the grid. Launch costs may also decline as more private enterprises enter the space-launch market. At some point, these factors may combine to greatly improve the economies of space-based solar power.

What can possibly make this affordable? Well, that all depends on the cost of energy and how much of a value we place on the environment. The cost of energy production is not as simple as dollars, Euros, or yen. What is the cost to the planet of the strip mining required for the coal we burn in our thousands of power plants? What is the payoff in reduced defense spending that will result in us not having to depend upon the volatile Middle East for oil to generate electrical power? How much is it worth to eliminate the acid rain associated with the burning of fossil fuels? What benefits will we reap from a power system that produces no greenhouse gases? The authors contend that when the real societal costs are considered, as well as the real monetary cost from end-to-end, space-based solar power begins to look like a winner. It is an expensive winner, but a good investment nonetheless.

A Possible Scenario

To examine the possible economic potential of space-based solar power, consider the possible decrease in launch costs if the Space Exploration Technologies (Space-X) Falcon-Heavy, formerly called the Falcon-9 Heavy is used to orbit a space-based solar power pilot plant. According to the Space-X website, this booster is built upon the successful Falcon-9, which began operational cargo-supply missions to the ISS in 2012. Much of the development costs have therefore been supported by NASA funding. When ready for its first scheduled flight (in 2013, according to Wikipedia), Falcon-Heavy will be capable of lifting 53,000 kg to low Earth orbit (LEO). According to V. Jaggard, Falcon-9 Heavy may initially reduce LEO launch costs to about $1,500 per kg.

We assume here that the payload of an operational Falcon-Heavy consists of a 2×10^4-kg space-based solar power pilot plant and the equipment necessary to deploy and transfer it to geosynchronous orbit (GEO). The total launch cost is estimated as $100 million.

To estimate the achievable near-term output of this power station, we referred to the literature for information regarding solar photovoltaic cell thickness and efficiency. According to a paper by B. O'Regan and M. Gratzel, experimental solar photovoltaic cell thickness was approaching 10 µm (10 millionths of a meter) as early as 1991. As A. Shah and colleagues have reported, experimental solar cell efficiencies were exceeding 20 % in 2000.

The following arguments about near-term profit possibilities from space-based solar power uses V. Jaggard's estimate that the typical U.S. household pays $0.11 per kWh of consumed electrical power. One dollar therefore supplies the typical U.S. household with about 300,000 J of electrical energy.

The thickness of the power station's array is conservatively assumed to be 20 µm, the cell efficiency is 0.2 (20 %), and the cell specific gravity is assumed to be 3 (which corresponds to a density of 3,000 kg/m^3). When the 2×10^4 kg power station is fully deployed in geostationary orbit, its area facing the Sun is about 300,000 m^2. For a square array, this corresponds to a side of about 580 m. Since the solar constant at the solar orbit of Earth is about 1,400 W/m^2, the power station receives about 466 mW of sunlight. About 20 % of this is converted into electricity in the solar array and 60 % of that electricity is beamed to Earth. The electrical power supplied to the grid is therefore about 56 mW. This could supply the baseline electrical needs of thousands of households.

It is not yet possible to estimate the projected operational lifetime of these hyperthin solar cells in the space environment. But since (as noted by V. Jaggard) the Japanese Space Agency (JAXA) hopes to launch and deploy a space-based solar-power pilot plant by 2030, several thin solar cells were attached to JAXA's experimental solar sail Ikaros, which was launched on an Earth-Venus trajectory in 2010. When questioned by author Matloff during a symposium in Aosta, Italy, during July 2011, Ikaros project director Y. Tsuda responded that no solar cell degradation was noticed during the first year in interplanetary space. It is therefore assumed that, like conventional thicker solar cells, the ones comprising this hypothetical pilot plant will have a lifetime of 30 years in the space environment.

During its 30-year operational lifetime, this electrical utility operating this power station delivers 5.3×10^{16} J to its customers. This converts to about 15-billion kWh. The projected income from this station is therefore about $1.6 billion. Launch costs are about $100 million. If development and infrastructure costs raise the total expenses to $600 million, the profit over a 30-year period approximates $1 billion.

Lots of things could spoil this projection. Development costs for the Falcon-Heavy could escalate, in-space life expectancy for hyperthin cells could be less than 30 years, etc. But at first glance, near-term development and implementation of such a space-based solar-power pilot plant does not seem unreasonable.

On-Demand Power: A Niche Application

In addition to supplementing the power grid, space-based solar power can be used to send power to localized areas that have an urgent need for electricity. As demonstrated by the aftereffects in the U.S. northeast after Superstorm Sandy and the nor'easter that followed it in late 2012, such needs are not imaginary. It is perhaps for these applications that space solar power will find its use due to the reduced overall power demand and the commensurate lower-mass systems that will be required.

Following a natural disaster, one of the first priorities is getting power into the affected area. While utility crews busy themselves in restoring power from the grid in a process that can take weeks or months, urgent care providers have an immediate need for electricity from the moment they arrive. Lives are often at stake. What if a constellation of relatively small orbital power satellites is placed in low to medium Earth orbit—much like the satellites in the Global Positioning System—so that one or more of them has line-of-sight power transmission capability to just about any point on Earth? Power generated by these satellites could be beamed to the affected area as soon as a ground station receiver is put in place.

As discussed by V. Jaggard, the process could be simplified and sped up if power were laser-beamed to a receiver attached to a stratospheric tethered balloon and then beamed down as microwaves to portable ground-level receivers. Application of space-based solar power to disaster relief efforts could reduce the long logistical convoys currently required to keep fuel flowing to the first responders, making them independent of the rest of the world for immediate power.

Further Reading

The National Space Society has an excellent online space solar power website: http://www.nss.
    org/settlement/ssp/. Another excellent resource is the "Space-Based Solar Power As an Opportunity for Strategic Security" Report to the National Security Space Office, October 10, 2007.
    We also recommend Peter E. Glazer, Frank P. Davidson, and Katinka Csigi, *Solar Power Satellites: A Space Energy System for Earth* (New York, Wiley, 1998).
Descriptions of research progress in the field of hyper-thin photovoltaic cells are presented by
    Brian O. Regan and Michael Gratzel, "A Low-Cost, High-Efficiency Solar Cell Based on Dye-Sensitized TiO_2 Films," *Nature*, 1991, 353: 737-740. Another good source is A. Shah, P. Torres, R. Tscharner, N. Wyrsch, and H. Keppner, "Photovoltaic Technology: The Case for Thin-Film Solar Cells," *Science*, 1999, 285, 692-698. A more recent web reference on the subject of space-based solar power is Victoria Jaggard, "Beam It Down: A Drive to Launch Space-Based Solar," *National Geographic News*, Dec. 5, 2011, http://news.
    nationalgeographic.com/news/energy/2011/12/111205-solar-power-from-space/   (accessed Nov. 20, 2012).

Chapter 13
Environmental Monitoring from Space

Growing Pains

Civilization spread across the planet.
The Neolithic gave way to the Bronze Age.
Three millennium ago, iron supplanted bronze.
Human populations expanded.
The papyrus soon gave way to the book.
The deductive philosopher was followed by the
    scientist.
As our numbers approach 7 billion,
our impact upon the world has become enormous.
From the depths of space, our instruments silently
    monitor environmental change.
What are we doing to our planet's atmosphere?
What are we doing to our planet's oceans?
What are we doing to other forms of terrestrial life?
Only now, do we begin to appreciate the cost of high
    civilization.

May 15, 2000   August 3, 2000

G. Matloff et al., *Harvesting Space for a Greener Earth*,
DOI 10.1007/978-1-4614-9426-3_13, © Springer Science+Business Media New York 2014

"The people know the salt of the sea
And the strength of the winds
Lashing the corners of the earth.
The people take the earth
As a tomb of rest and a cradle of hope.
Who else speaks for the Family of Man?
They are in tune and step
With constellations of universal law."
—From the poem "The People Will Live On" by Carl
Sandburg

In the late 1990s, the author Frank White visited Huntsville, Alabama, and lectured at the local university on what he called the "overview effect." White had just authored a book by that title and was touring the country speaking on the subject and promoting his book. According to White, most of the world's astronauts experience an epiphany when they first gaze back from space upon this wonderful cradle of life we call Earth. With its magnificent blue oceans, swirling cloud formations, and obvious land formations, these space explorers experience a sense of belonging to the globe, not necessarily as citizens of the United States, Russia, or whatever country was responsible for their making the journey, but as human citizens of planet Earth.

From space, without the human-created boundaries dividing the landforms, there is no obvious distinction between the United States and Mexico, between Israel and Syria, or between any other bordering states. Those who have experienced the overview effect claim to be forever altered in their perspective regarding not only countries but also their peoples and the planet.

Many who have not been to space can experience this vicariously by looking at the famous "Earth Rise" photograph taken by the Apollo 8 astronauts during their journey to the Moon. Shown in Fig. 13.1, this image is credited by many as awakening their awareness of how apparently fragile this lonely jewel of life may be among an uncaring and deadly cosmos.

Other than an ethereal and somewhat elusive moment of spiritual awakening and often-needed perspective broadening, of what use can this and other photographs be for monitoring and protecting the planet? A large scientific community that is engaged in the field of remote sensing provides the answer to this question.

Remote sensing is the term used to describe virtually any method of viewing something from a distance, as opposed to being physically present wherever the measurement is being made. Satellite remote sensing describes the functional ability to detect electromagnetic radiation from Earth's surface or atmosphere and to understand what the observed radiation's characteristics mean regarding the conditions from wherever they originate.

This chapter describes how our ability to access space helps us understand, predict, and prepare for local and global environmental change using various remote sensing techniques and technologies.

Fig. 13.1 Earth, as seen by the crew of Apollo 8 (courtesy of NASA)

Visual Images

Since we humans first ventured into space with either our robotic emissaries or in person, we have been taking pictures. In fact, we have taken lots and lots of pictures. As a NASA employee, author Les Johnson is fortunate to engage in regular mission debriefings with astronauts when they return from space, having been aboard the space shuttle or the International Space Station. Their debriefings are always in the form of a slide show, reminiscent of one created to describe a family's recent trip to the beach, showing all aspects of the mission from preflight preparations to touchdown. A large number of their photographs are of Earth from space. Some of the photos are taken with nothing more than a high-grade commercially available camera; others are taken with high-resolution cameras designed for that very purpose. Whatever their source, they show the various landforms on Earth, and often pay particular attention to whatever geographic region the astronaut photographer calls home. In addition, over the half-century of spaceflight, these pictures tell an evolving environmental story.

The United Nations recently released a compilation of such pictures in a book entitled *One Planet Many People*: *Atlas of Our Changing Environment*

(New York: United Nations Environment Program, 2005). One of the many striking images from the compilation shows what happened to unprotected land near the Iguazu National Park in Argentina between 1973 and 2003 (Fig. 13.2).

Spring Comes Earlier Each Year

According to the April 1, 2008, issue of *Science Daily*, spring is arriving about 5 days earlier to Eurasian forests than it used to. Using satellite data, researchers were able to track the appearance of leaves as a function of time over several spring seasons, showing an unmistakable trend toward spring arriving progressively earlier each year. After examining two decades of data, scientists learned that not only is spring arriving earlier than it used to, but also that fall is becoming tardy, arriving 10 days later than it did just two decades ago. A similar trend is seen in data collected over North America.

Will an earlier spring and later fall result in a more productive growing season for farmers? Or will it usher in a longer dry season with more extreme temperatures? The verdict is still out, but one thing is for sure—with satellite monitoring, we will be able to understand what is happening as never before.

Biodiversity and Transgenic Crop Monitoring

Genetic engineering and space technology at first thought might not be considered to have much in common, other than owing their existence as fields of research to the scientific revolution. But it is possible that one may enable careful monitoring of the other to ensure compliance with environmental regulations and to project future food crop health and availability.

In recent years, taking advantage of the biotechnology revolution, agriculture companies are developing food crops that are genetically modified to resist certain pests. One of the first commercial successes in this area is an insect-resistant strain of corn called "Bt corn," which is genetically engineered corn containing bacterial genes that express an insecticidal protein from *Bacillus thuringiensis*. In theory, use of Bt corn will increase crop yield due to its not being consumed as readily by insects. Proponents also believe Bt corn will benefit the environment because fewer pesticides will be required; if the food crop is inherently bug resistant, why would a farmer need to use as much pesticide?

Rather than allow wholesale adoption of this new strain of corn, farmers are planting Bt as only a portion of their overall crop. They are also, for the most part, isolating it from their regular corn crop. From the farmer's point of view, this makes good sense. After all, it is a new product, and what if it does not work as advertised? From an environmentally conscious consumer's point of view, it is also good to

Fig. 13.2 These pictures, taken in 1973 and 2003, respectively, show how civilization has consumed forestland in Paraguay and Brazil. The Iguazu National Park in Argentina, which is protected from development, remained intact (courtesy of the United Nations Environmental Program)

understand how this corn might crossbreed with regular corn and, perhaps, spread beyond the confines of where it was initially planted.

The U.S. Environmental Protection Agency (EPA) would like to monitor these crops for a variety of reasons and to answer some very pertinent questions:

- Is the Bt corn actually as pest resistant as advertised?
- Will Bt corn spread or crossbreed with regular corn? Alternatively, can it be geographically controlled?
- Are farmers planting this new corn as planned? Or are some cheating and planting more than stated?

There are not enough field agents to monitor the 25 million acres of corn planted across Middle America. What is needed is a way to monitor large areas remotely, preferably from the air or space.

The EPA is working with NASA to determine if the light reflected from genetically modified Bt corn is sufficiently different so as to allow it to be observed from space using a technique called hyperspectral imaging. When the human eye or a scientific instrument observes an object, the observation is actually of the light reflected from the object that reaches the eye or the instrument. In the case of most vegetation, we observe green light. Although the human eye can only distinguish various shades of green, our scientific instruments can detect many more reflected colors. Each different shade or color corresponds to a different wavelength of light. When these multiple colors are examined simultaneously, they are considered hyperspectral. In the case of the observations of interest to the EPA and NASA, approximately 120 colors are observed simultaneously.

If this as-yet-unproven technology is viable, it will provide a large-scale monitoring capability to inform the grower and governmental regulatory agencies of potential pest resistance development in the Bt corn, its ability to be geographically contained, and exactly which crops entering the food supply are transgenic. Flying regularly overhead, hyperspectral instruments aboard spacecraft may provide the data needed. In addition, this capability will provide decision makers with vital data on the annual health of food crops as climate shifts and both crops and pests adapt to their changing environmental niches.

Global Sea Levels

Approximately 70 % of Earth's surface is covered by water. Another statistic that is not as well known, but is very pertinent as we face the very real possibility that global sea levels will rise significantly in the next century, is that over half of the world's population lives within 100 km of an ocean. If sea levels do rise, then a large percentage of the total population will be affected.

In the recent geologic past, sea levels have fluctuated a great deal. During the last ice age, which occurred about 20,000 years ago, sea levels were about 120 m lower than today. This happened because an enormous amount of water was contained in

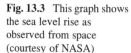

Fig. 13.3 This graph shows the sea level rise as observed from space (courtesy of NASA)

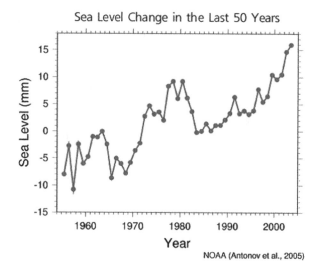

massive ice sheets that covered parts of North America and northern Europe. By contrast, during the previous interglacial period, about 125,000 years ago, the sea level was at least 4 m higher than today.

Global warming can produce a rise in sea level through the thermal expansion of seawater and the net melting of glacial ice. We are careful to use the word *net* when describing the melting of sea ice. What counts is how much more ice melts than forms in any given time period (see below).

Ground measurement of sea level changes due to environmental effects is not straightforward. The following events can affect any given sea level measurement:

- The tides (cyclical and predictable)
- Atmospheric pressure (how hard the atmosphere pushes down upon the water beneath it)
- Winds (particularly during storms)
- Changes in ocean currents (where large volumes of water flow at any given time)
- Seasonal changes in density (due to temperature changes and the amount of salt dissolved in the water)
- The relative height of the measuring device (Is the land sinking due to erosion?)

The TOPEX/Poseidon and Jason 1 spacecraft have been using radar to monitor sea levels since 1992. Satellite data, combined with that from tide gauges (Fig. 13.3) shows that Earth's mean sea level has increased almost 15 cm over the last 50 years. When looking at the graph, it is important to note that the satellite data are different from what was recorded in ground-based tide gauges. The possible causes for the discrepancy range from geologic changes in gauge height to satellite instrument calibration errors. Regardless of the validity of the absolute value of the changes, the trend line is what is of most interest, and that trend is toward increasing ocean sea levels.

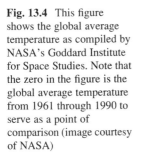

Fig. 13.4 This figure shows the global average temperature as compiled by NASA's Goddard Institute for Space Studies. Note that the zero in the figure is the global average temperature from 1961 through 1990 to serve as a point of comparison (image courtesy of NASA)

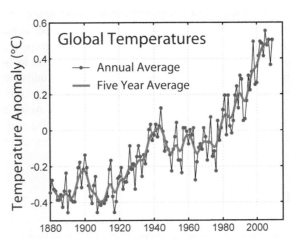

Global Ice Measurements

Much of the world's freshwater is locked in glacial ice at either the South Pole or in the Greenland ice sheet. If the net amount of ice in these glaciers changes, so will the level of the world's oceans. Satellite observations of the Greenland ice sheet show an increase in the so-called melt zone along its edge. This region is still mostly ice, but an increasing portion of its surface area is melting during the summer months, sending rivers of water flowing into the sea.

Satellite observations also show that the thickness of the ice in Antarctica is increasing, taking water from the sea and locking it away in increasingly thick ice near the South Pole. An open question is whether or not the rate at which the ice is getting thicker at the South Pole will balance the amount melting in Greenland and in the Arctic. The net difference will help us understand the potential contribution to the recorded sea level rise from changing environmental conditions at Earth's poles.

Atmospheric Temperature Monitoring from Space

Spacecraft are now used to monitor the temperature of the atmosphere at various altitudes. They do not directly measure the temperature, as they must fly above the atmosphere in the vacuum of space. Instead they observe changes in other atmospheric parameters and infer (calculate) the temperatures at various altitudes. The atmospheric temperature profiles obtained depend on the technique used to make the calculation, and there are different technical approaches that often produce somewhat different results.

The satellite-based evidence of global warming is convincing (Fig. 13.4).

Localized heating is also observable from space. "Urban heat islands" have been observed for years, highlighting the localized impact from human activity on temperatures in our cities. It is no surprise that the temperature in our cities tends to be higher than in vegetative regions. One has to simply feel the heat radiating from an asphalt street on a mid-summer's day and compare it to the relatively cool grass in the neighborhood part to intuitively understand why.

It was a surprise, however, when NASA's Aqua and Terra satellites observed that the nighttime surface temperature under wind mill farms tended to be higher by almost a full degree than that of nearby terrain without windmills.

Desertification

The transformation of once-vegetative or agricultural land into desert is called desertification. Often directly brought about by human activity, and significantly influenced by climate change, large portions of once-arable land and inland lakes are being transformed into deserts. Activities attributed to cause this burgeoning loss of useful land include urban development, overcultivation, poor irrigation practices, overgrazing, and overt destruction of vegetation. Examples of overt destruction would include clear cutting of forests without regard to reforestation or burning of forests for land cultivation, often resulting in only a few successful growing seasons on the recently cleared land before it is subsequently depleted.

Whenever productive or potentially productive agricultural land becomes desert and unusable, a local scarcity of food is likely to develop. In some locations, particularly in developing countries with limited resources and mobility, the results can be catastrophic. In the 1970s and 1980s, the Sahara Desert grew in size, displaced local and formerly self-sufficient populations, and turned them into refugees.

In the early 1970s, the United States launched a series of Landsat satellites that began inventorying land use across the globe. By looking at the data from Landsat and other, more recent, satellites, it is possible to piece together a picture of how much land has been lost to desertification over time. The left side of Fig. 13.5 shows an image of Chad Lake in Africa taken from space in 1972. The right side of the figure is an image of the same portion of Chad Lake taken in 2001. When you compare the pictures, you can see that a once water-filled lake has been turned into a marsh, overgrown with vegetation. If the water levels do not soon begin to rise, and the water loss trend continues, this lake will soon be dry. And what of the farmers and human population nearby that depend on the lake water? They will potentially become a displaced, migratory population.

Lest one think that desertification is a problem only for the developing world or just in Africa, as a resident of the southeastern United States, author Les Johnson has to look only a few hundred kilometers to the east to see a similar problem occurring in Georgia, near the growing city of Atlanta. Atlanta uses water from

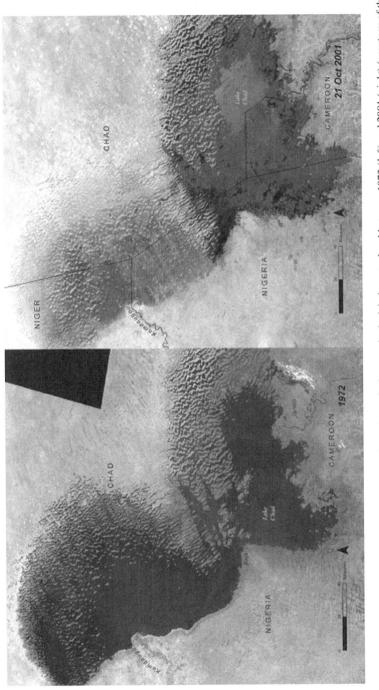

Fig. 13.5 Satellite images of Africa's Chad Lake showing a dramatic decrease in the lake's water level between 1972 (*left*) and 2001 (*right*) (courtesy of the United Nations Environmental Program)

Fig. 13.6 Volcanic
eruptions are on such a huge
scale that they can be easily
seen from space (image
courtesy of NASA)

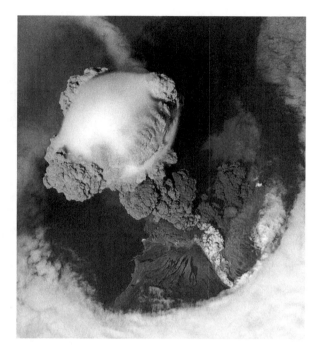

Lake Lanier to supply much of its needs. The problem is that Lake Lanier is running
dry due to the severe drought that has gripped the region for several years. With a
major urban population pulling water from the lake and nature not providing
sufficient rainfall to replenish it, the lake's water level is dropping dramatically.
At this writing, there is disagreement among the states of Tennessee, Alabama, and
Georgia over who owns the water in nearby rivers and how that water will be
allocated. The debate has been acrimonious—and this is within the borders of a
prosperous and stable country. Imagine the issues to be resolved should these have
been separate countries!

Natural Disasters

Earthquakes, tsunamis, hurricanes, tornados, floods, and wildfires. Nature continues to humble we humans on a scale unimaginable to the individual on the
ground as the disaster is experienced. As technology advanced, people began to
appreciate more and more the strength and destructive power of nature as they
were able to traverse the distances impacted by one disaster or another. With
satellite imagery, the true scope of a hurricane several hundred miles in diameter
can be seen—not to mention the devastating path of a tornado half a mile wide and
on the ground for 100 miles, as is all-too-often experienced by people in the
middle parts of North America.

Fig. 13.7 Minot, North Dakota's, Souris River flooded the city in May 2011. Landsat 5 (*top* picture) and 7 (*bottom*) spacecraft took these "before" and "after" pictures of the city showing how much area was flooded (image courtesy of NASA and the U.S. Geological Survey)

Several commercial companies sell this imagery to governments and the public. Fortunately, an incredible amount is available for free thanks to the astronauts aboard the International Space Station. For example, in 2009 the crew of the ISS captured an image of Sarychev Peak volcano in Russia's Kuri Islands (Fig. 13.6). Figure 13.7 shows the devastation wrought on Minot, North Dakota, when the Souris River overflowed its banks in May 2011. Not only are these images powerful in conveying the destructive power of nature, but they can aid policymakers in deciding where to locate critical infrastructure (such as power plants, etc.).

On a more local level, disaster management teams can use near-real time satellite imagery to save lives and protect property. Several wildfires started in southern California in early May 2013. Figure 13.8 shows a fire started near Camarillo Springs, California, as the prevailing winds blew it toward the coast, taking smoke out over the Pacific Ocean.

Fig. 13.8 NASA's Terra satellite took this image of a California wildfire in 2013 (image courtesy of NASA)

Conclusion

Monitoring the global environment can best be accomplished by using the unique vantage point of space for remote sensing in addition to comprehensive ground and aircraft observations. When taken together, the data will provide scientists with a better understanding of our planet and its changing climate and environmental conditions. Continued observations will also allow us to understand how much progress we are making as we alter the way humans interact with Earth and march toward a sustainable future in which the direction of environmental change is not toward degradation but toward regeneration and renewal.

Further Reading

Frank White's book, *The Overview Effect,* is a mind-expanding experience that is highly recommended reading. For more information about the U. S. Environmental Protection Agency's interest in monitoring the growth and spread of transgenic crops, see an excellent summary article in the September 11, 2003, issue of *Nature* titled, "U. S. Reflects on Flying Eye for Transgenic Crops." A sad but educational "must read" is the U. N. publication, *One Planet Many People: Atlas of Our Changing Environment* (New York: United Nations Environment Program, 2005).

Chapter 14
Protecting Earth

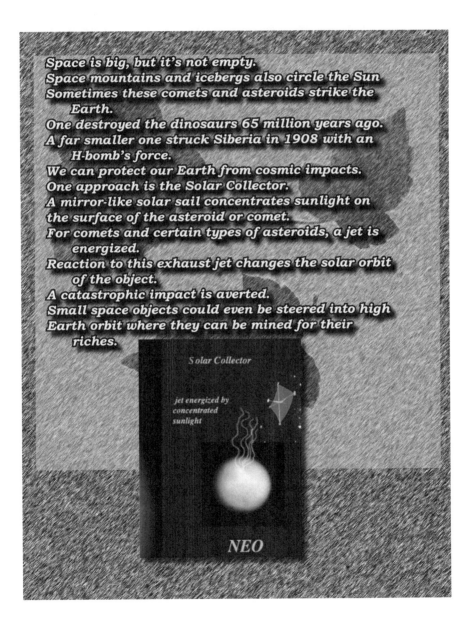

Space is big, but it's not empty.
Space mountains and icebergs also circle the Sun
Sometimes these comets and asteroids strike the
    Earth.
One destroyed the dinosaurs 65 million years ago.
A far smaller one struck Siberia in 1908 with an
    H-bomb's force.
We can protect our Earth from cosmic impacts.
One approach is the Solar Collector.
A mirror-like solar sail concentrates sunlight on
the surface of the asteroid or comet.
For comets and certain types of asteroids, a jet is
    energized.
Reaction to this exhaust jet changes the solar orbit
    of the object.
A catastrophic impact is averted.
Small space objects could even be steered into high
Earth orbit where they can be mined for their
    riches.

Solar Collector

jet energized by
concentrated
sunlight

NEO

G. Matloff et al., *Harvesting Space for a Greener Earth*,
DOI 10.1007/978-1-4614-9426-3_14, © Springer Science+Business Media New York 2014

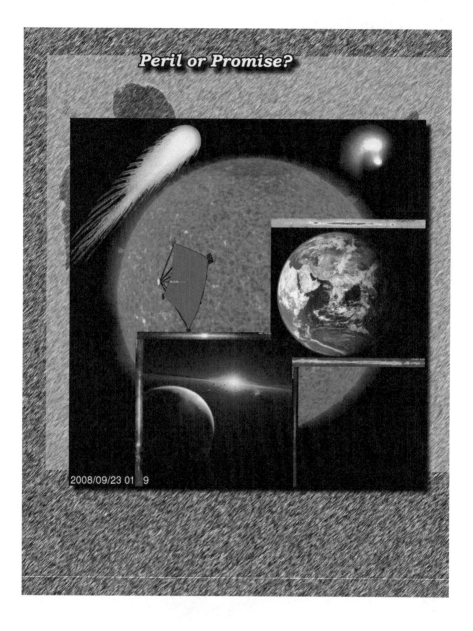

2008/09/23 01 9

> *"Full fathom five thy father lies:*
> *Of his bones are coral made;*
> *Those are pearls that were his eyes;*
> *Nothing of him that doth fade*
> *But doth suffer a sea change*
> *Into something rich and strange."*
> —From the play *The Tempest* by William Shakespeare

Intense tempests of celestial origin have blown through the skies of Earth, obliterating landscapes and sending towering tsunami through the oceans. Such events have extinguished vast numbers of living organisms; some of these die-offs have been altered by geological processes into fossils.

For the first time, it is almost in our power to prevent or alleviate these events. But the action of protecting Earth from certain types of cosmic catastrophes will change us into something "rich and strange": a spacefaring species.

Catastrophes: Terrestrial and Celestial

There are many types of catastrophes that have threatened Earth life, or may threaten Earth life in the future. Some are of terrestrial origin, some of celestial origin. Only some can we alleviate with our advanced technologies.

One terrestrial catastrophic event that may have resulted in mass extinctions in life's long history is the super volcano. It is not impossible that tectonic forces have on rare occasions produced a volcanic eruption dwarfing Vesuvius and Krakatoa as the atomic bomb dwarfs a hand grenade. If something like that occurred today, civilization (and perhaps humanity) would be doomed by the long-duration obscuration of sunlight that would put Earth in a freezer. We cannot predict such an event, and we have no way of preventing it—though perhaps a few representatives of humanity and other species might survive in underground shelters or aboard space habitats.

However, super volcanoes, as destructive as they are, are a purely local phenomenon. There is no way that a terrestrial volcano can affect hypothetical biospheres elsewhere in the Solar System, or on planets circling other stars.

Stellar explosions are one class of non-local cosmic catastrophes. When a star considerably more massive than the Sun reaches the end of its hydrogen-burning life, stellar fires dim, and the star collapses upon itself. But the collapse increases temperature and density in the star's interior, which allows a host of energy-releasing thermonuclear reactions involving elements more massive than hydrogen and helium.

In a cosmic instant, the dying star becomes a supernova, converting perhaps 1 % of its mass into energy and outshining most of the stars in its galaxy for a few days or weeks. A significant fraction of the energy emitted by a supernova is in the form of gamma rays and X-rays. In the case of solar emissions, Earth's upper atmosphere shields us from harmful radiation. But the intense blast of radiation from a nearby supernova might overwhelm this atmospheric shield, extinguishing most life on our planet's surface.

A nearby supernova explosion would produce very bad effects. Once again, if we had sufficient warning, our best option would be to dig a deep hole.

It is possible that a nearby supernovae caused some of the mass extinctions in the fossil record. Perhaps subterranean life forms then had the opportunity to re-colonize our planet's surface.

Happily, most massive stars approaching the supernova stage are very distant from our Solar System. One candidate near-future supernova progenitor with a mass of 100 Suns is Eta Carinae. Happily, this unstable giant resides at a comfortable distance of 7,500 light years (1 light year is approximately 63,000 times greater than the average Earth-Sun separation, or 63,000 astronomical units).

A typical supernova might be bad for life on planets within a few hundred light years. But there are rare cosmic events—so-called gamma ray bursts—that are even more powerful. These have been observed in other galaxies using gamma-ray telescopes in Earth orbit and may be due to the collapse of super-massive stellar cores into spinning black holes. Happily, there is no evidence to date that such monsters lurk dangerously close to our Solar System.

There are no technological insurance policies to protect us from galactic catastrophes such as nearby supernovae or gamma ray bursts. But we have reached the point at which we might be able to do something about the danger posed by asteroid and comet impacts.

Menace from the Skies

Most icy comets currently reside in the distant, frigid Oort Cloud,[1] spending most of their existence tens of thousands of astronomical units from the Sun. These remnants of the Solar System's formation would spend eternity in the far reaches of space if it were not for the fact that stars, including our Sun, revolve around the center of the galaxy. Our Sun requires about 250 million years to complete one circuit through the Milky Way Galaxy.

Once every 100,000 years or so, a Sun-like star makes a random close approach to our Solar System, passing through the outer fringes of the Oort Cloud. On these occasions, the solar orbits of some comets will be altered. A fraction of the affected comets will be flung out of the Solar System; others will be directed towards the inner Solar System, perturbed into highly elliptical paths requiring tens of thousands of years to complete one orbit around the Sun.

---

[1] The Oort Cloud is a spherical cloud of comets believed to orbit the Sun at a distance of about 50,000 astronomical units.

Fig. 14.1 Asteroid Ida appears to be one large piece of rock (image courtesy of NASA)

As they enter the inner Solar System, solar heating causes gas and dust to evaporate from the 10- to 20-km nucleus of one of these long-period comets. A spherical coma consisting of evaporated water, ammonia, and methane gas, with dimensions of perhaps 10,000 km surrounds the tiny nucleus. Solar radiation pressure pushes one or more tails from the coma, each perhaps 100 million kilometers in length; comet tails are always directed away from the Sun.

If one of these comets targets Earth, it will approach from deep space or near-solar space at around 40 km/s. Warning time of the impending collision will be quite limited.

Space is big. Earth impacts by long-period comets are not frequent. But there are other sky objects to be concerned about. Another source of potentially dangerous icy objects is the Kuiper Belt, a region of icy, comet-like dwarf planets including Pluto, extending from about 30 to 50 astronomical units from the Sun. The larger of the Kuiper Belt objects (KBOs) are in relatively stable orbits, never entering the inner Solar System. But collisions and giant-planet alignments sometimes produce fragments or alter orbits. Then, small KBOs can enter the inner Solar System as short-period comets, comets with orbital periods measured in years or decades. A few of these have perihelia (closest approaches to the Sun) within Earth's orbit. Because of their varying orbital inclinations, meaning that their orbits do not lie within the same plane as Earth's orbit, most known KBOs will never threaten our home planet.

Rocky and stony space objects, the asteroids, are generally located closer to the Sun than the comets and KBOs. Most of them reside in the Asteroid Belt, which extends from about 2 to 5 astronomical units from the Sun. The smallest of the asteroids are about the size of boulders; the largest approximate the size of Texas. To make matters more complicated, there are at least two types of asteroids—those that appear to be solid rock (Fig. 14.1) and those that are essentially "rubble piles" held loosely together by their weak gravitational attraction (Fig. 14.2).

Fig. 14.2 A close-up photograph of asteroid Itokawa, which appears to be made of many smaller pieces of rock (courtesy of the Japanese Aerospace Exploration Agency)

However, collisions and giant-planet perturbations have altered the orbits of some of these. Several thousand near-Earth objects (NEOs) between about 20 m and 40 km in size pass close by Earth. Some of these have the spectral signatures of rocky and stony asteroids; others are most likely extinct comets.

Most NEOs that have been observed to date have been detected by government-funded programs. But private organizations are beginning to get into the NEO-detection and tracking act. The B612 Foundation, for example, which is named for the asteroid in Antoine de St. Exupery's novel *The Little Prince*, seeks funds to establish a dedicated in-space telescope capable of accurately tracking 500,000 asteroids that occasionally approach Earth. Apollo astronaut Rusty Schweickart and retired NASA astronaut Ed Lu are involved in the effort.

Although the threat of Earth-impacting comets cannot be completely ignored, it is the NEO population that represents the greatest threat. But we are beginning to observe them systematically and will soon have the capability to deflect Earth-threatening NEOs.

Defending Earth from NEO Collisions

It may have been a 10-km NEO that contributed to the demise of the dinosaurs 65 million years ago. Smaller NEOs in the 100-m size range pack a smaller wallop, but one of them could still destroy a city. And there are a lot more of the smaller ones.

Potential city-killing NEOs strike Earth at intervals of about a century. The last known impacts were in 1908 and 2013—both in sparsely populated regions of Russia.

According to the NEO observers, an object dubbed Apophis is scheduled to make a very close pass of Earth in 2029, perhaps approaching within 35,000 km.

Although the chances of this few-hundred meter-wide object colliding with Earth in 2029 are minimal, it will be back in 2036. If Apophis is an extinct comet, gravitational tides caused by Earth may produce a tail during the close approach. And if the reaction to the hot gases emitted by the tail is just right (perhaps we should say just wrong), Apophis's solar orbit could be very slightly altered so as to put it on a collision course with Earth on its return 7 years later. If the impact of a NEO of Apophis's size occurred on land, a state or small country would be obliterated. If it occurred at sea, the resulting tsunami would drown millions.

There are advantages and disadvantages of some techniques proposed to divert Earth-threatening NEOs. It is assumed in this discussion that the capability will exist to deliver spacecraft in the 100,000-kg mass range, either human-crewed or robotic, to the vicinity of the offending NEO. Many national or international space programs including those of the United States, Europe, Russia, Japan, China, and India are developing large interplanetary spacecraft to become operational starting in 2015–2020. From the point of view of Earth protection from threatening NEOs, this development is none too soon.

If the Earth-threatening object is from the Oort Cloud, collision warning time will be minimal. In such a case, terrestrial space agencies would probably elect to take the most dramatic action. Earth-launched spacecraft would most likely be robotic and their payloads would probably consist of what are euphemistically referred to as nuclear or thermonuclear devices.

These 100- or 1,000-megaton explosives would be ignited as close to the approaching object as possible, in the hopes of altering the object's trajectory. But there are problems with this nuclear option.

One problem is geopolitical. A Saturn-V class booster capable of delivering nuclear weapons to the approaching comet might be viewed by some nations as the world's most capable ballistic missile. If it were armed with many 1-megaton H-bombs instead of one huge explosive, it could serve as a weapon capable of destroying several major cities with a single launch. So planning for the venture would have to be both transparent and international.

However, there is a more basic scientific issue with the nuclear option—at least for some comets. It simply might not work.

One class of comets in highly elliptical orbits, the so-called "sungrazers," literally kiss the Sun's visible surface during their very close perihelion passes. Some of these have been observed to "calve," or fragment at perihelion.

Another example of a comet breaking into many fragments is Shoemaker-Levy 9. In July 1992, this comet approached Jupiter within 100,000 km. Shoemaker-Levy 9's trajectory was altered by the giant planet's gravity field so that the comet returned to Jupiter on a collision course 2 years later. Astronomers observing the collision of the comet with Jupiter were fascinated to observe that prior to the collision, the comet fragmented into many smaller objects (Fig. 14.3). The likely cause of this fragmentation was the added pull of Jupiter's very strong gravity acting non-uniformly on the comet.

The nuclear option should only be applied as a last resort. Instead of deflecting or destroying an approaching comet, a large nuclear bomb may well cause it to break

Fig. 14.3 A Hubble Space Telescope composite image showing fragments of Comet Shoemaker-Levy 9 approaching Jupiter on a collision course (image courtesy of NASA)

into many radioactive fragments, each of them still targeting our planet. The end effect would be akin to trading death by a bullet fired from a handgun to death by a close-range shotgun blast. In both cases you are quite dead.

A somewhat less dramatic approach is kinetic deflection. Earth orbits the Sun at 30 km/s. NEOs in approximately circular orbits will have about the same solar orbital velocity.

Let's say we launch a robotic spacecraft equipped with a low-thrust propulsion system, such as an ion drive or a solar sail. If the orbit of the spacecraft is gradually changed so that it is traveling in the opposite direction to Earth and most NEOs, its relative velocity to an Earth-threatening NEO will be around 60 km/s.

Timing and guidance must be very precise. But if the spacecraft is directed at this relative velocity to impact the NEO, it will pack quite a wallop. According to some model calculations, the momentum change delivered to the NEO might alter the trajectory enough to prevent a collision with Earth, if the warning time is sufficient. Of course, some classes of NEOs may fragment rather than altering course—which would not be a good thing!

Several Chinese scientists have proposed such a kinetic option in case Apophis's second approach threatens Earth. Shengping Gong and colleagues calculate that the high-velocity impact of a small solar sail on Apophis in 2028 could avert a 2036 impact.

Solar Options and the Gravity Tractor

If collision warning time is measured in years or decades and if NEO mass, trajectory, and other characteristics are sufficiently well known, there are several less dramatic options. These use forces of nature—solar radiation pressure and gravity—to deflect an offending NEO without explosives or impacts

One of these benign approaches is to encircle the Earth-threatening NEO with a reflective solar "parasol." To visualize this approach, imagine that an NEO is like a potato. Astronauts land on the "potato" and stick a number of toothpick-like structures into it. These toothpicks support a spherical, highly reflective, thin-film solar sail, which can be modeled by a sheet of aluminum foil.

What you've done is to increase the model NEO's reflectivity to sunlight and its surface area. In the real world, the effects of these changes would be to render the NEO's solar orbit more elliptical. If the warning time is measured in decades, such an orbital alteration might be enough to change a direct hit on Earth into a near miss.

An innovative modification of this approach was proposed by MIT graduate student Sung Wook Paek in 2012. Paek describes a scenario in which a robotic spacecraft approaches an Earth-threatening NEO and bombards the object with a series of reflective "paint balls." The increased reflectivity of the object would result in greater solar radiation pressure that would alter the NEO's solar orbit slightly over a period of many years or decades.

Another solar approach is the "solar collector," which is shown schematically in Fig. 14.4. The solar collector is essentially a two-sail solar sail. The collector sail faces into the Sun and focuses sunlight on the smaller thruster sail. This thruster directs a concentrated sunbeam on the NEO's surface. If the NEO is coated with layers of dust, soil, or ice, a jet of superheated material (like a comet's tail) may be raised in the direction of the thruster sail. The reaction force to this jet pushes the NEO in the opposite direction (downward in the figure).

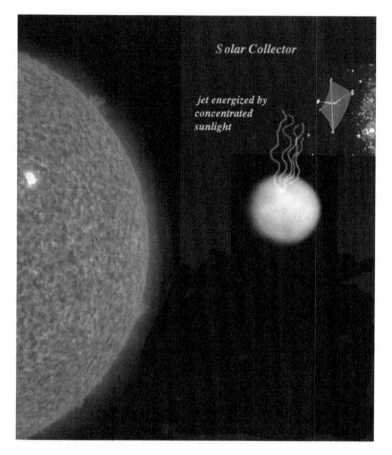

Fig. 14.4 The solar collector in operation near a NEO

For 100-m NEOs and solar collectors in the 50–100-m size range, the solar collector seems capable of deflecting NEOs to prevent Earth collisions if warning times are a year or more. But there are issues with maintaining the solar collector on station near the NEO for long time periods and protecting the thruster and collector surfaces from damage by the escaping jet. But the solar collector is the only NEO deflection option that allows for the possibility of steering an NEO into high-Earth orbit where it could be mined for its resources. For more about asteroid mining, see Chap. 11.

The Planetary Society has funded a study of a solar-collector variant, the laser collector. Called Laser Bees, this concept is being investigated by the universities of Glasgow and Strathclyde in Scotland. The laser collector replaces the sunlight concentrator with a sun-pumped laser. It has the advantage of a smaller "hotspot" on the NEO surface but requires more moving parts and radiator panels to eliminate waste heat.

All of the deflection options discussed so far have the disadvantage that they can be applied to only certain types of NEOs. None of them would be effective in the case of a rubble-pile NEO, for example. A very interesting recent concept that works conceptually with any type of NEO is the gravity tractor.

Imagine that a large spacecraft equipped with a low-thrust propulsion system (possibly an ion drive) flies in formation with the Earth-threatening NEO, at as close a distance as safely possible. If the thruster is non-operational, the small gravitational attraction of the NEO will pull the spacecraft towards it until the two objects ultimately collide. But if the thruster is used to maintain a constant separation between NEO and spacecraft, the spacecraft's smaller gravitational pull slightly alters the NEO's solar trajectory. Given enough collision warning time and accurate knowledge regarding the NEO's trajectory, an Earth collision could again be converted to a near miss.

Human Missions to NEOs

Before we can perfect our NEO-defection techniques, we must learn a great deal about these objects. By 2020, humanity's interplanetary capabilities should be sufficiently mature that we can begin to conduct NEO exploration missions.

At least for the nearest NEOs, round-trip travel times (including 1–2 week stopovers at the NEO) should approximate 90 days. Although NEO exploration will be more risky than roundtrips to the Moon, weightlessness and cosmic radiation should be less problematical than in 2–3 year roundtrips to Mars. Within a few decades, humans will almost certainly have begun the exploration of these small celestial objects.

Further Reading

To check on the current state of Eta Carinae, consult Francis Reddy, "The Supernova Next Door," *Astronomy*, 2007; 35 (6): 33-37. The possible connection between gamma-ray bursts and black holes is discussed by Steve Nadis, "The Secret Lives of Black Holes," *Astronomy;* 2007, 35 (11): 29-33.

Various astronomical data sets present the orbits and properties of asteroids and comets. One of these is Arthur N. Cox, ed., *Allen's Astrophysical Quantities*, 4th ed., (New York: Springer-Verlag, 2000).

To survey craters produced on Earth by past asteroid and comet impacts, see Francis Reddy, "Illustrated: Earth Impacts at a Glance," *Astronomy*, 2008; 36 (1): 60-61. The possible problem posed by Apophis's visit in 2026 is reviewed in Bill Cooke, "Fatal Attraction," *Astronomy*, 2006: 34 (5): 46-51.

Many astronomy texts discuss the interaction of comet Shoemaker-Levy 9 with Jupiter. One is Eric Chaisson and Steve McMillan, *Astronomy Today*, 6th. ed. (Upper Saddle River, NJ: Prentice-Hall, 2008) (1999).

An excellent source to review many of the NEO-deflection concepts is T. Gehrels, ed., *Hazards Due to Comets and Asteroids*, (Tuscon, AZ: University of Arizona Press, 1994). The parasol concept was introduced in the paper by Gregory L. Matloff: "Applying International Space Station (ISS) and Solar-Sail Technology to the Exploration and Diversion of Small, Dark, Near Earth Objects (NEOs)," *Acta Astronautica*, **44**, 151-158 (1999). For current model calculations regarding the solar collector, see Gregory L. Matloff, "The Solar Collector and Near-Earth Object Deflection," *Acta Astronautica* , Volume 62, Issues 4-4, February-March 2008, pp. 334-337 and "Deflecting Earth-Threatening asteroids Using the Solar Collector," *Acta Astronautica*, 2012..

The gravity tractor is such a new concept that little has appeared as yet in refereed publications. A web article discussing this concept is by Russell Schweickart, Clark Chapman, Dan Durda, Piet Hut, "Threat Mitigation; The Gravity Tractor, (White Paper 042)" http://arxiv.org/pdf/physics/0608157.pdf (accessed Nov. 25, 2012).

NASA and other space agencies have begun planning human missions to NEOs. See, for example, Andre Bormanis, "Worlds Beyond," *The Planetary Report,* 2007; 27 (6): 4-10.

A paper describing Chinese research on kinetic NEO interception is available on-line. It can be accessed as Shengping Gong, Junfeng Li, and Xiangyuan Zeng, "Utilization of H-Reversal Trajectory of Solar Sail for Asteroid Diversion," http: arxiv.org/ftp/arxiv/papers/1108.3183.pdf (accessed Nov. 20, 2012).

The B612 Foundation and other groups investigating NEO detection and diversion are described in Bruce Lieberman, "Asteroid Watch," *Air & Space Smithsonian*, 2012; 27 (6), 62-65. For a description of Laser Bees, see Bruce Betts, Jr., "Zapping Rocks with Lasers: Saving the World by Destroying Space Invaders," *The Planetary Report*, 2012, 32 (2), 20-21.

Sung Wong Paek's "paintball" approach to NEO diversion is discussed in a number of on-line sources. One, in the Oct. 26, 2012, on-line issue of the *Los Angeles Times* is Thomas H. Maugh, Jr., "Asteroids Headed to Earth Could Be Diverted with Paintballs," http://www.latimes.com/news/science/sciencenow/la-sci-sn-asteroid-paintballs-20121026,0,2746770.story (accessed Nov. 25, 2012).

Chapter 15
Mitigating Global Warming Using Planetary Engineering

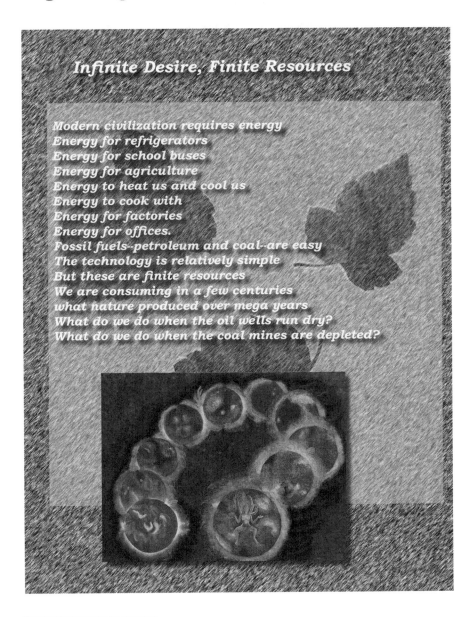

Infinite Desire, Finite Resources

Modern civilization requires energy
Energy for refrigerators
Energy for school buses
Energy for agriculture
Energy to heat us and cool us
Energy to cook with
Energy for factories
Energy for offices.
Fossil fuels--petroleum and coal-are easy
The technology is relatively simple
But these are finite resources
We are consuming in a few centuries
what nature produced over mega years
What do we do when the oil wells run dry?
What do we do when the coal mines are depleted?

(Adapted from an article by Robert G. Kennedy III, PE, Kenneth I. Roy, PE, Eric Hughes, David E. Fields, Ph.D.)

G. Matloff et al., *Harvesting Space for a Greener Earth*,
DOI 10.1007/978-1-4614-9426-3_15, © Springer Science+Business Media New York 2014

"We saw Thee in Thy balmy nest.
Young dawn of our eternal day!
We saw Thine eyes break from their east,
And chase the trembling shades away."
From the poem "The Shepherd's Hymn" by Richard
Crashaw

It is a strange time on Planet Earth. For the first time in humanity's recorded history, we no longer can feel totally secure in Mother Nature's "balmy nest." There are simply too many of us, and we all desire to live well. As far as we know, humanity is the first terrestrial species with the technological power to alter Earth's global environment in a measurable way. We seem to be in a tight fix of our own making. Can we do anything about it?

This is not the first time that our little home world has experienced climate change. In fact, it can be argued that climate change has actually been a fact of terrestrial existence since the origin of our Solar System. As is the case for all stable main sequence stars, our Sun's luminous output has been gradually increasing as it ages on the main sequence. This process began almost 5 billion years ago, and it will continue until the Sun leaves the main sequence and devours the inner planets (including possibly our Earth) as it expands to its giant phase more than 5 billion years in the future.

Early in the Solar System's history, Earth would have become a frigid, perhaps lifeless, ice ball like Mars if it were not for greenhouse gases deposited in our infant planet's atmosphere by impacting comets. Greenhouse gases are a big problem today, but 4 billion years ago or so they trapped absorbed radiant energy from our then sub-luminous star. They may have been absolutely necessary to increase terrestrial temperature so that primitive life could develop, thrive and evolve into more complex organisms such as ourselves.

At any stage of its development, a planetary surface can be treated as a thermo-dynamic system. The light that warms our little world is generated as high-energy photons in the thermonuclear furnace deep within the Sun. As it percolates slowly towards the Sun's visible surface (or photosphere) on its multi-million-year journey, it is gradually downshifted in energy towards the (mostly) visible light we see. After reaching the photosphere, it requires a bare 8 min to complete a 93 million mile (150 million kilometer) journey to Earth's solar orbit.

A very tiny fraction of the Sun's radiant output strikes our planet. Much of this energy is scattered by Earth's atmosphere or reflected back into space by Earth's land and ocean surface. The light that is absorbed is later re-emitted as invisible infrared radiation.

If this were the entire story, Earth would be as cold and devoid of surface life as Mars. Ice would cover our planet's oceans. Desolate winds would howl across the empty deserts.

What maintains Earth's moderate temperatures are the greenhouse gases in our planet's atmosphere, water vapor (H_2O) and carbon dioxide (CO_2). Essentially, these raise the planet's average temperature. Before the advent of industrial technology, the levels of these and other greenhouse gases in Earth's atmosphere were fairly constant or slowly varying with time. But the technological break-throughs beginning in the nineteenth century have greatly altered this balance.

In the natural state, there is a balance. Large-scale tectonic events such as volcanoes spew large amounts of fine dust particles (called aerosols) into the upper troposphere and lower stratosphere. Aerosols reside for a sufficiently long time in the upper atmosphere to increase Earth's reflectivity of sunlight and thereby reduce the planet's global temperature.

If greenhouse gases get the upper hand in Earth's natural state, global warming occurs. This results in shrinkage of the polar ice caps, higher sea levels, expansion of deserts and a greater frequency of violent storms of tropical origin. If atmospheric aerosols predominate, Earth cools and an ice age results as polar caps expand.

Although the present episode of global warming has a lot to do with individual and collective human actions, the root causes developed in deep time. More than 400 million years ago, during the Devonian Period, life had engaged in the large-scale colonization of the land. For the first time, extensive forests covered much of the planet's surface area. But the climate gradually cooled. Episodes of tectonic activity—volcanoes and earthquakes—buried the remains of many early forests. Gradually, this organic material was compacted into peat, coal, and petroleum. Even after more than a century of industrialization, Earth's reserves of these fossil fuels are enormous. But our civilization is essentially consuming in a few centuries a planetary inventory of carbon deposited over a period of perhaps 100 million years.

As industrial or technological processes combine the hydrocarbons stored in fossil fuels with atmospheric oxygen during the combustion process, a lot of energy is produced, which is a good thing. But an unavoidable consequence of hydrocarbon combustion is the release into the atmosphere of water vapor and carbon dioxide.

Although these are both greenhouse gases, the time water vapor stays in the atmosphere is limited, since more water vapor usually results in more precipitation. But the increased level of carbon dioxide in the atmosphere is not a good thing, since this greenhouse gas remains in the atmosphere for a longer period of time. Given enough time, of course, biological activity (photosynthesis by plants) will convert much of the excess CO_2 into oxygen. But the big question is, how much time is required for such a correction? Painstaking observations and analysis have revealed that there is almost certainly a direct correlation between the global increase in atmospheric CO_2 levels, the rate of fossil fuel consumption and global temperature increase. So far at least, naturally occurring photosynthesis does not appear to be rapid enough to correct for the excess CO_2.

So the question remains—what can we do about it? Ultimately, chemical engineers and chemists may come up with efficient and inexpensive methods of trapping or sequestering CO_2 generated by fossil fuel consumption at the source. In time, green power technologies such as wind or solar will heavily supplement or even replace fossil fuels. The next chapter deals with in-space methods of slightly reducing the amount of sunlight striking the planet and thereby reducing global temperatures at a controlled rate. The balance of this chapter considers pros and cons of geo-engineering. Can we safely increase the planet's solar reflectivity or photosynthesis rate using Earth-based technologies and thereby reduce global warming in a controlled fashion?

Volcano Simulation: Artificial Aerosols in the Stratosphere

One obvious suggestion is to chemically produce large quantities of aerosol particles and release them, possibly from high-altitude balloons, in the upper atmosphere. This would duplicate on a global scale the local effects of many volcanic eruptions.

In principle, this seems very easy. Volcanoes produce a lot of ash and dust, and they fling this material to high altitudes. Figure 15.1 presents a photograph of a recent eruption of the Pavlov volcano. This photo was taken by astronauts aboard the International Space Station on May 18, 2013. Pavlov is located in the Aleutian Island chain, about 625 miles (1,000 km) southwest of Anchorage, Alaska. This eruption spewed volcanic ash and dust to an altitude of about 20,000 ft (6,000 m).

So why not engineer many artificial volcanoes, simultaneously injecting many tons of fine particles into the upper air all over the globe? Such an activity would certainly lower global temperature.

Not so fast! As with many apparently easy solutions to global warming, there are problems with this approach. The residence time of aerosol particles is a complex combination of aerosol composition and size and atmospheric parameters including temperature and wind structure and seasonal variations. It is very difficult to accurately predict how long a given aerosol particle will remain aloft and where it will finally descend.

If human civilization adopts a policy of global-scale high-altitude artificial aerosol-injection to alleviate global warming, the goal of moderate global temperature reduction may well be overshot. It is not impossible that our planet may swing from one extreme to the other and enter a global ice age. Also, we know very little about the interaction between the descending aerosol particles and local and global ecologies. These effects may not be pretty.

Local Reforestation and Artificial Phytoplankton Blooms

A somewhat more benign approach is to encourage the planting of many trees to partially replace those lost in the Amazon basin to agricultural development and those lost globally to desertification. If enough trees can be planted in the developed world to more than compensate for forest loss in the developing world,

Fig. 15.1 A volcano eruption imaged from space (courtesy of NASA)

photosynthesis by these large plants could remove a significant fraction of the CO_2 injected into the atmosphere by fossil fuel consumption.

A number of cities have announced the laudable goal of planting a million or more trees in what constitutes urban forests. According to Wikipedia, participating cities include Los Angeles, New York, Shanghai, Denver, London and Ontario.

Another suggested approach to increasing global photosynthesis is to fertilize the oceans with iron particles to increase the growth of plankton and other oceanic phytoplankton species. The rationale behind this concept is that small iron particles are a significant nutrient for these microscopic oceanic plant species that contribute significantly to global photosynthesis.

Unfortunately, small-scale experiments in 2004 performed by the Woods Hole Oceanographic Institution, located in Massachusetts, revealed a number of problems that might prevent large-scale implementation of this approach. One obstacle is that the amounts of atmospheric carbon dioxide removed by natural algal blooms do not appear large enough to alleviate global warming. Field studies in the southern ocean between Antarctica and New Zealand indicate that our knowledge of deep-ocean circulation is limited. As the Woods Hole press release points out, the oceans already remove roughly one-third of the CO_2 injected into the atmosphere by human activities.

Another problem is that natural or pollution-enhanced algal blooms, such as the one imaged from space in Fig. 15.2, often are damaging to the local ecology. Imaged by NASA's Landsat-5 on October 5, 2011, this algal bloom in Lake Erie was likely fertilized by industrial pollution or sewage. The perhaps greater potential ecological damage caused by global-scale artificial algal blooms must weigh heavily in the minds of environmental decision makers.

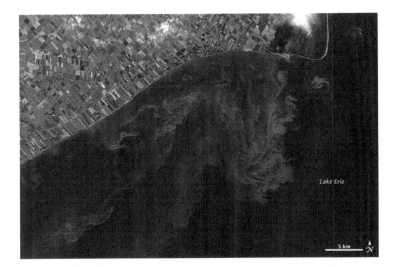

Fig. 15.2 Algal bloom in Lake Erie imaged from space (courtesy of NASA)

Increasing Earth's Reflectivity

Another suggested geo-engineering solution to the problem is a global program to increase the fraction of incident sunlight reflected back to space by Earth, called the planet's albedo.

One interesting possibility requires action by individual homeowners and municipal governments, not national governments. More and more of the world's population is becoming concentrated in urban environments. So what would happen if all or most urban homeowners simply decided to paint their roofs white?

An interesting paper co-authored by Hashem Akbari and two other Canadians investigates this possibility. They conducted a series of computer simulations for settings in Earth's temperate zones. If urban roofs were generally painted white and light-colored pavement was utilized, an urban area's albedo might increase by about 0.1. Global cooling might reach 0.01–0.07 K, which corresponds to an equivalent CO_2 emissions reduction of 25–150 billion tons.

Although the idea certainly has merit, here are obvious issues regarding global dependence on such an approach. First, how would such a procedure be enforced on an international scale? What are the incentives for individuals and local governments who would implement such a strategy, and what would be the penalties for those who refuse to comply? Also, how do the cost and energy requirements for producing light-colored and dark-colored roofing material and pavement material compare? Finally, de-centralized solar energy requires in many cases high-efficiency rooftop solar panels that absorb rather than reflect sunlight. Would we do better in terms of atmospheric CO_2 reduction to encourage decentralized solar energy or albedo-enhancing techniques?

Conclusions: Limits and Alternatives

This chapter considers only a few of the suggested techniques for correcting or partially alleviating global warming using geo-engineering. For a more comprehensive treatment, refer to Appendix 3 of this volume, which is authored by Robert Kennedy III, Kenneth Roy, Eric Hughes and David Fields.

Large-scale global geo-engineering schemes may have severe ecological consequences and be difficult to stop before the global-pendulum swings towards a new ice age. Small scale, local approaches such as reforestation are laudable and should be encouraged. But there may be consequences if urban albedo-modification is applied on a large scale.

The next chapter returns therefore to the main theme of this book. It discusses a suggestion to utilize mined asteroid material to create a huge, stationary sunshade in space. Properly designed and positioned, such a device could reduce the solar energy incident on Earth and consequently partially or entirely alleviate global warming. And if global cooling seemed too rapid, it would be an easy task to move the structure from its stationary location between Earth and the Sun.

Further Reading

There are many sources that describe the formation of fossil fuels from early terrestrial forests. One very readable source is John Reader, *The Rise of Life*, Knopf, NY (1986).

If you would like technical information on the interaction of solar electromagnetic radiation with natural aerosol particles in Earth's atmosphere, an excellent reference is K. Ya. Kondratyev, *Radiation in the Atmosphere*, Academic Press, NY (1966). Many books deal with parameters controlling natural and pollutant aerosol atmospheric residence time – one is Laurent Hodges, *Environmental Pollution*, 2nd ed., Holt, Rinehart, and Winston, NY (1977).

You can access online descriptions of experiments with oceanic iron fertilization. The Wood's Hole source is a news release dated April 16, 2004, with the URL http://web.archive.org/web/20061231181525/http://www.whoi.edu/mr/pr.do?id=886 (accessed Nov. 22, 2013).

Research on urban albedo modification is summarized by Hashem Akbari, H. Damon Matthews, and Donny Seto in "The Long-Term Effects of Increasing the Albedo of Urban Areas," *Environmental Research Letters*, Volume 7, Number 2 (2012), which is available online at: http://iopscience.iop.org/1748-9326/7/2/024004/article (accessed Nov. 23, 2013).

Chapter 16
Mitigating Global Warming Using Space-Based Approaches

Climate change is real.
We may argue about how much is caused by
  humans
  and how much is caused by nature.
But we are not helpless.
One option is the Sunscreen.
Since we may move asteroids to prevent Earth
  impacts
Could we disassemble one and locate it
  appropriately
  to block some of the sunlight striking Earth?
We might reduce erosion in this way
We might save the polar caps
But will people never change their ways?
If a technology is developed to reduce global
  warming??

General layout of sunshades near the Sun-Earth L1 region

(Adapted from an article by Robert G. Kennedy III, PE, Kenneth I. Roy, PE, Eric Hughes, David E. Fields, Ph.D.)

All in a hot and copper sky,
The bloody Sun, at noon,
Right up above the mast did stand,
No bigger than the Moon
—From the poem "The Rhyme of the Ancient Mariner" by
Samuel Taylor Coleridge

Conservation, recycling, more efficient machines, altered lifestyles and new sources of energy are all needed to reduce the growth rate of our greenhouse gas emissions. Unfortunately, the best all of these actions combined can achieve is a reduced rate of emission growth. If humans are truly causing global warming by profligate use of fossil fuels, then we need to do far more than reduce the growth in our emissions. We must reduce the absolute amount of these gases in the atmosphere to levels below those of the last century. With more and more people entering the global middle class, and their commensurate use of more and more energy, it is unlikely, perhaps even impossible, that we will be able to stop global warming. We almost certainly will not be able to conserve our way back to pre-twentieth century atmospheric carbon dioxide levels. We should do all that is necessary to slow down the rate at which we dump CO_2 into the atmosphere, but we are kidding ourselves if we believe we will be able to actually reduce the amount already within it. Conservation, recycling, more efficient machines and altered lifestyles are all band aids for a problem that requires reconstructive surgery.

If we cannot eliminate the thing that is causing global warming, then is there anything we can do to mitigate it? The answer is "yes."

There are two elements to global warming. The first, and the one about which most people concentrate their efforts, is the greenhouse gas element. Sunlight warms our planet, and the greenhouse gases in the atmosphere trap some of the heat from the solar energy, keeping the overall planet warm (even at night) and making it possible for life to exist. Without a greenhouse effect, Earth would be a very cold place. Humanity's efforts are now focused on reducing the amount of heat trapped in our atmosphere by these gases. But what about the other part of the story? What if we could reduce the amount of heat generated? What if we could reduce the overall amount of sunlight hitting Earth so that global atmospheric temperatures can return to what we consider "normal"? What if we were to build a large sunshade in space and reduce the amount of sunlight striking Earth, thus allowing it to cool?

Giving Earth Sunglasses

Geo-engineering, or planetary engineering, is the application of technology for the purpose of influencing the properties of a planet on a global scale. This chapter considers space-based geo-engineering techniques, as opposed to the planet-based techniques discussed in the previous chapter. Large sunshades positioned between Earth and the Sun can be used to heat or cool our planet. Such an arrangement can cope with global cooling or warming here on Earth or to modify the global properties of other planets to make them more Earth-like. Sunshades can also displace energy generation from an overstressed biosphere, yielding substantial economic benefits in the form of avoided costs, revenue streams, and new capabilities.

Gravity keeps your feet on the ground, satellites in orbit and the Moon in its orbit around Earth. Gravity also keeps Earth in orbit around the Sun. The Sun's immense mass is pulling on Earth, and if it weren't for our motion in orbit around the Sun, we would certainly fall into it. But Earth's mass also pulls on the Sun.

There is a point in space at which the pull of Earth's gravity exactly equals that of the Sun pulling upon Earth. This point is much closer to Earth than to the Sun due to our much smaller mass and is called the Sun-Earth Lagrange point, or L1. It is called a Lagrange point in honor of the mathematician who worked out the formulae describing the physics used to describe it. In the eighteenth century, the Italian mathematician Giuseppe-Luigi, Count of Lagrange, finalized the mathematical theory that describes the behavior of celestial objects at the so-called Lagrange points.

If you want to see the Sun-Earth L1 point, go outside whenever it's daytime and look up (for a moment only!) at the center of the Sun. It's right there, only some 1,500,000 km away. Compare this distance with that of Earth's Moon, which is about 400,000 km away, or just one-fourth the distance. But, while the Moon moves against the background of fixed stars over a period of a month, the Sun-Earth L1 point will always be directly between Earth and the Sun. Something placed there will obstruct some of the sunlight that would otherwise hit Earth. This action is also known as "occultation," which has nothing to do with magic, though it may look like it. Note that this effect is not an eclipse either, because in general the dark part of a shadow—known as the umbra—from an object at L1 cannot reach Earth. Unless the object at the Sun-Earth L1 is very large—thousands of kilometers in diameter—only the dim outer part of the shadow (the penumbra) reaches Earth. Placing enough of something at the Sun-Earth L1 can significantly cut down the sunlight hitting Earth. If we're clever, we can locate just the right amount of "stuff" to put there to alter the solar radiation hitting Earth to achieve any desired temperature reduction. (Or, with a similar technique, an increase, because sometimes ice ages happen, too.) Also, these techniques can in principle be applied to any other planet.

Another important characteristic to note is that although the Lagrangians are called "points," they are in fact *regions* of stability. For example, the region of metastability centered around the Sun-Earth L1 point above is shaped like a sausage, lying about 800,000 km along the orbit (i.e., perpendicular to this page). The region's transverse dimensions (i.e., up and down, and left to right on this page) are approximately 200,000 km each. Thus the Sun-Earth L1 region contains roughly 30 quadrillion cubic kilometers of space, which is of a similar order to the 200 quadrillion cubic kilometers of cislunar space in the Earth-Moon system.

However, for purposes of shading sunlight on Earth, the "sweet spot" is a lot smaller, less than 1 % of the region.

Finally, not all shade is created equal. From the diagram you can see that a sunshade parked right on the Sun-Earth line intercepts a higher quality of light than one parked somewhat off the axis. This phenomenon, which varies by a factor of 2.5 or so, is called solar limb darkening, and is one of the components of shading efficiency. (You can observe this phenomenon for yourself in an ordinary low-wattage incandescent light bulb. Note how the edge of the bulb looks dimmer than the center no matter what your vantage point.)

Sunshades in Space

If we are principally interested in reducing by some degree the amount of solar radiation hitting a planet and doing so with as little effort as possible, then the "stuff" we position between a planet and its star will need to have maximum surface area for minimum mass. It will need to block a lot of light and not be too heavy to launch into space. Fortunately, a technology is being developed today that might be up to the job: solar sails.

A solar sail is just what its name implies. It is a sail that propels itself using sunlight. Sunlight has no rest mass, but it does have momentum. When light is reflected from an object, the light imparts some of its momentum to that object—just like the cue ball hitting the four ball in a game of billiards. The momentum of sunlight is very, very small. So small, in fact, that we cannot feel it, and it is negligible on Earth, where the magnitudes of all the other forces acting on us are so much larger. Once you get into space, without air and wind and in situations where the pull of gravity from Earth or the Sun is also small, the tiny push from sunlight can be a significant factor in making a spacecraft move. If the spacecraft has a large, lightweight and highly reflective sail attached, it can maneuver just about anywhere in the inner Solar System without fuel—using only reflected sunlight to propel it.

However, these sunshades won't be solar sails in the traditional sense. There will be some big differences. Solar sails for space-based geo-engineering can be very heavy compared to solar sails used to transport cargo around the Solar System. They won't have a payload, in the sense that they will be the payload. Also, because they won't be hauling designated freight, we don't have to make them any particular size. We can make them any size we want, and we will make that choice based on ease of manufacture and perhaps other factors. For our purposes we don't really care if we have a million little sails each a kilometer square or a single large sail having an area of a million square kilometers. The optical properties of the sails are critical, and will vary a lot depending on the particular engineering solution chosen. They won't necessarily be totally reflective mirrors, or totally absorptive blackbodies. Different parts may be mirrored, or black, or diffractive, or even transparent—or some combination. Rather than lasting just for the duration of a cargo mission, they will be built for longevity. They will be engineered to endure the harsh conditions of interplanetary space for an extended period of time. They will have to withstand high radiation fields and continual assaults by the solar wind or even the occasional solar storm, and tolerate the occasional puncture by micrometeoroids.

Further, our sails will have to have sensors and controls and even some degree of intelligence. In order to remain precisely at the L1 point, they will have to vary the thrust resulting from solar radiation to counter forces that would pull them off station. Because we are talking about many thousands or possibly millions of such solar sails cruising along in orbit in close proximity to each other (much like a giant school of fish), they will also have to be social. Their sensors will observe their neighbors as well as the primary and the satellite, and they must be able to maneuver to avoid crashes or other conflicts, such as cutting off a neighbor's light. They must have means to allow them to receive instructions from their makers. We may want them to move out of position if temperatures on the planet drop too far. Not only would these sails block light like a parasol, they could use some of the light that they intercept for free stationkeeping thrust, rather than conventional reaction mass squirted from an engine.

Sunlight in space is effectively free, unlike rocket fuel shipped up from Earth or somewhere else, which would get expensive over the long run. The sails could easily generate enough electrical power for their onboard computers and other equipment via photovoltaics on their bright side. There is no reason they might not generate a surplus over their own modest requirements for power, either; perhaps far more.

To distinguish these rather specialized solar sails from the ones discussed in other chapters we propose the term "Dyson Dot." This is a deliberate allusion to the original idea of the Dyson sphere, which was proposed by Freeman Dyson as a system of orbiting space facilities designed to completely encompass a star, thus capturing its entire energy output. The Dyson Dot would be designed to block or capture only limited amounts of a star's radiation that would otherwise strike one of its planets.

A typical solar sail needs to achieve a mass to area ratio on the order of $10\text{--}20 \text{ g/m}^2$ to be useful. Our Dyson Dots can be that light if we like or they can be much heavier. There is no reason they couldn't range up into the kilogram per square meter range if necessary. One factor to keep in mind is that lighter Dyson Dots will experience a significant acceleration due to light pressure, and this requires that they be positioned somewhat closer to the Sun away from the traditional L1 point—perhaps 500,000 km closer to the Sun than the traditional L1 point. If it were heavier or less reflective it could be closer to the usual L1 point.

I stopped here. To understand why this is so we have to go back to the physics of the L1 point. The planet's mass produces an attraction on an object at L1 that is exactly opposite to the gravitational attraction produced by the Sun—in effect reducing the observed mass of the Sun allowing the object at L1 to have a solar-orbital period exactly equal to that of the planet. But with a solar sail we now have a third force component to deal with and that is the force resulting from sunlight hitting and to a greater or lesser extent being reflected from the solar sail or

Dyson Dot. This third force adds to the force of the planet's gravity on the Dot—in effect moving the point of stability sunward from the usual L1 point.

Dyson Dots as a Solution to Earth's Global Warming Problem: How Much is Enough?

So, we can position a school of Dyson Dots at the Sun-Earth L1 point, or perhaps slightly sunward of it, but how much total area do we need? How much sunlight do we need to block? This is a simple question that doesn't have a simple answer. Depending on future energy policies, and on various global and solar cycles, we may need to either artificially cool or warm our planet. To begin to bound the problem, it is worth noting that a 0.25 % reduction in the Sun's energy output is what is estimated to have caused the Maunder Minimum. For unknown reasons, the sunspot cycle shut down between the mid sixteenth and seventeenth centuries. Astronomers refer to this period as the Maunder Minimum. Historians call it the Little Ice Age. During this period, the Thames River in England froze for the first time in recorded history, crops failed, population growth stalled, and sea ice cut of Iceland from Europe. The famous Danish astronomer Tycho Brahe recorded winter temperatures 2.7° Fahrenheit below average during the last two decades of the sixteenth century.

If we shoot for a similar reduction we're probably in the ballpark of what would be necessary to deal with global warming. It is important to note that while this approach could adjust the average global temperature, it would in no way address the other environmental issues associated with the continued burning of hydrocarbons.

Figure 16.1 illustrates why no stable sunshade can project a shadow spot of exactly the right size, namely no bigger than Earth's diameter. Some shading is unavoidably wasted due to the geometry of the Sun-Earth-L1 system. This is the other component of shading efficiency, and the highest possible value—about 82 %—is obtained right at L1, as depicted in the figure. However, as previously discussed, any sunshade made of a non-magical material will have to cruise somewhat inside of the Sun-Earth L1. The brighter the mirror, the further inside L1 it has to go, and paradoxically the heavier the overall mass and cost, and the less efficient the total project gets. Any real solution will be an exercise in optimizing and trading off against multiple constraints.

At the time of this writing, it appears that the gross parameters of an array of sails will vary by a factor of roughly 3: from 300,000 km^2 (about the size of the state of Arizona) and 10–20 million tons at the low end, to about 900,000 km^2 (somewhat greater than the entire West Coast of the U.S.) and 50–60 million tons at the upper end. For example, using reasonable middle values for the parasol parameters—80 % reflectivity or albedo, mass 53 g/m^2, positioned 2,100,000 km from Earth—we would need almost 700,000 km^2 of sunshade area to achieve a reduction of 0.25 % in the solar constant. That's some 37 million metric tons. (This sounds like a

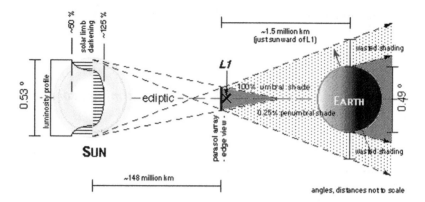

Fig. 16.1 General layout of sunshades near Sun-Earth L1 region

lot, but for perspective bear in mind that the United States alone burns about *one billion* tons of coal *every year*. One supertanker of the many hauling petroleum around the world's oceans weighs about half a million tons fully loaded; and 37 megatons is roughly just three days' supply of crude oil.) If each Dyson Dot has an area of 10 km^2 then our array, or school, would consist of 70,000 units.

As the mirror gets denser—say, 100 g/m^2 instead of 50—or the climate change problem becomes more severe—the total mass, hence the cost, needed to achieve the desired effect on Earth goes up. If we want a 1 % reduction in insolation, then the area, mass, and cost go up by a factor of 4; for 2 %, they go up by 8.

On another level in the grand scheme of things, 37 million tons isn't so much: it's the mass of a small stony-iron asteroid a mere 300 m across—a class of rock so small that we haven't seriously looked for yet.

Infrastructure Discussion

L1 is very high up there beyond this deep gravity well we live in, and our technology is grossly unequal to the task of lofting large solar sails into space. At a current launch cost of $10,000 per pound just to get into low Earth orbit (LEO), it is obvious that existing rockets are too expensive to implement this particular solution. Only a truly advanced spacefaring civilization using energy on an unprecedented scale could execute such a project. However, all proposed solutions to global warming—including the Do Nothing Plan that is always an option in human affairs—are expensive and/or painful. For example, it has been estimated that the cost to reduce CO_2 emissions just for the United States to 80 % of the 1990 level could cost up to 6 trillion dollars in today's money (2007). The cost to bring Chinese and Indian populations up to the US standard of living while lowering their current CO_2 emissions could easily be an order of magnitude more. Perhaps, becoming an advanced spacefaring civilization would be the cheaper alternative.

Several proposals have been presented to implement the Dyson Dots. Build a large manufacturing facility on the Moon to manufacture the Dots using lunar materials and then launch them from there. Build a large rail gun on Earth that could fire a spacecraft to the Sun-Earth L1 point and then deploy the Dots from the rocket. Move an asteroid to the Sun-Earth L1 point and then establish a manufacturing facility there to make and launch the Dyson Dots using material from the asteroid itself as the raw material.

Wherever it would be located, the manufacturing facility would be busy for some time. Using the previous medium-valued example, and assuming a deployment period of 10 years, we would need to launch about 10,000 tons of Dot stuff a day. Even after we had achieved the desired reduction in sunlight, we would still have to replace units as they fail.

Energy Options

Each square kilometer at Earth orbit receives some 1,400 MW of power. For the task of cooling the Earth, it doesn't matter how much solar radiation the Dyson Dot reflects just as long as it is prevented from getting here. If we convert the 20 % absorbed energy to electricity at 10 % efficiency we have 28 MW of power per square kilometer of mirror. In fact, solar cells—not to be confused with solar sails—have achieved efficiencies much higher than 10 % with current technology. If we absorbed 70 % of this energy instead of 20 %, and converted 20 % of that to electricity to be beamed back to Earth via maser (microwave laser), then we have almost 200 MW/km^2. Let's stick with the conservative 28-MW number. Keep in mind that we've got over 700,000 km^2 of solar sail somewhat sunward from the Sun-Earth L1 point. Assuming that a typical nuclear power plant produces 1,000 MW then our Dyson dot array can boast a power output equal to 20,000 nuclear power plants. The United States currently has a mere 100 or so such plants; they produce 20 % of the country's electricity. It is clear that if we could transmit even a small fraction of the sunpower which Dyson Dots intercept to Earth in the form of electricity, we would have enough to provide for all the nations of this world with plenty left over. Certainly displacing polluting power generation offworld would be a Good Thing, and a net benefit to the ecosystem, since terrestrial carbon burners waste two-thirds of their fuel's energy due to basic thermodynamic inefficiency.

The solar wind is composed mainly of hydrogen and helium. Some of the helium is the isotope He-3. Our mirrors will be exposed to the solar wind for many years. They could be designed to capture these particles and then return to a central processing facility at the end of their life span for recycling. Hydrogen, helium, and especially helium-3 could be very valuable to a space faring civilization. Helium-3 holds great promise as a fuel for fusion reactors that can be easily transported to Earth in an extremely compact hence valuable form.

Looking Backward: Aesthetics and Agriculture

Suppose at some time in the future that we've built a major manufacturing facility on the Moon and it has been building and launching mirrors for 10 years. Furthermore, suppose that the scientists and engineers then tell us we've reached our goal and have finally positioned enough mirrors at the proper point to reduce global temperatures to UN-specified levels. It's a world holiday of course. You go outside and look up at the Sun, using the appropriate eye safety gear provided. What do you see? Nothing special. It's not possible to detect even 700,000 km^2 of Dyson Dots floating somewhere inside the Sun-Earth L1 point against the backdrop of the Sun itself, because their dark umbras never get to you on the ground. Nor are you likely to feel the effect on your skin, any more than you are of the shadow of a bird flying overhead. Using special devices designed to image sunspots, you might see a strange cluster of black dots in the center of the sun if you were willing to make the effort. To most people going about their normal affairs on this planet, however, they would be invisible, like all good infrastructures. You shrug and decide to go seek more interesting sights.

The farmers in this future have pointed out at the start of the Dyson Dot that a quarter-percent reduction in sunlight means a quarter-percent loss in crop yield. Yet these folks, no less obstreperous than the ones today, have been assured that no such shortfall would happen. It turns out that plants are not very efficient at using the sunlight provided to them. Chlorophyll uses only very narrow bands of wavelengths of light in the blue-purple and orange-red regions of the spectrum. They paradoxically don't use green at all, which is right near the *peak* (most abundant energy) of the solar spectrum. Green light is reflected away, which is why green plants look that way. So a number of Dyson Dots have been modified to augment groundside agriculture by converting some of the energy they collect to particular frequencies used by chlorophyll. This supplemental light is beamed to Earth using weak lasers carried on special Dots. There is even talk of augmenting these frequencies above natural levels, but only for the Earth's agricultural regions. Best of all to most voters, including you, is the continuing dividend that the Dyson Dot Consortium has been paying to the citizens of Earth via power sales and reduced insurance premiums. (The economic justification is a topic for another book, however.)

A Bigger Perspective

Earlier we mentioned other worlds, and other places. For example, if we wanted to reduce the solar radiation hitting our neighbor Venus to Earth-normal levels, as opposed to the venuforming of Terra which seems to be going on right now, we would need to block about half (48 % exactly) of the incoming sunlight. This can also be done using Dyson Dots located at the Sun-Venus L1 point, but the level of effort would be approximately 200 times as great as that required to address the global warming problem right here on Earth. We recognize that the terraforming of Venus would involve much more than just adjusting its solar constant, but that task would have to be part of the overall terraforming effort. If we were terraforming Mars on the other hand, we would do just the inverse—double the Martian insolation with off-axis Dots. Projects like this could take thousands of years without quasi-magic methods like self-replicating nanotechnology. Nevertheless, how much would a second Earth in this Solar System be worth to the human race? Survival trumps the ordinary calculus of economics.

Given the Anthropic Principle, none of these considerations are unique to our home system. Physical laws are the same for everybody; thus it is reasonable to suppose that any intelligent beings may utilize techniques we would recognize. Dyson Dots as described in this chapter can be used to increase or decrease the solar constant and to some extent modify the color of light hitting the target planet. They can be used as power stations and as resource collection vehicles. If a race were interested in terraforming a planet, this would be a handy technique to employ. If global warming is a real threat can we deploy this tool in time to make a difference? Maybe. Certainly not without cheap and reliable access to space. And certainly not without the national or international will to do great things. It is clear that building and using Dyson Dots would create a set of mutually reinforcing capabilities that would each be valuable, even indispensable, to a spacefaring civilization. Solving the climate change problem here on Earth may be just the thing to bootstrap the human race to a new level of existence.

But keep in mind that although we may find it too daunting at present, other races may already have employed this technology. Perhaps SETI researchers should begin to look for the occasional flash from distant Dyson Dots. They're designed to reflect a lot of energy and it must occasionally be seen by outside observers. They may not want to talk to us, but it would somehow be comforting to know that some other race is making itself comfortable on a distant world.

Further Reading

The guest authors Ken Roy, Robert Kennedy, and David Fields are all affiliated with the Department of Energy Oak Ridge National Laboratory in Tennessee. Their biographies are in the Appendices. Since they completed this chapter as Chapter 14 of the First Edition of this book, they have presented this concept at a number of scientific and technical meetings and published related papers in refereed journals. See for example Robert Kennedy III, Kenneth I. Roy and David E. Fields, "Dyson Dots: Changing the Solar Wind to a Variable with Photovoltaic Solar Sails" *Acta Astronautica*, Vol. **82**, pp. 225-237 (2012). This paper was originally presented at the Seventh IAA Symposium on Realistic Near-Term Advanced Scientific Space Missions in Aosta, Italy on 11-13 July, 2011 with the title "Dyson Dots." For additional information on this topic, consult Appendix 4 of this book, which is co-authored by these authors, and Eric Hughes.

For additional information on the topics discussed in this chapter, we recommend the following:

Fogg, M. J., *Terraforming: Engineering Planetary Environments,* Society of Automotive Engineers (1995).

Friedman, Louis, *Starsailing: Solar Sails and Interstellar Travel,* Wiley (1988).

Glaser, Peter, *Solar Power Satellites,* Arthur D. Little Inc. (1968).

Hayden, Thomas, "Curtain Call" in *Astronomy* magazine, edited by B.B. Gordon, January 2000, pp. 45-49.

Kraft, Christopher C., *The Solar Power Satellite Concept,* NASA-JSC-#14898, US GPO 1979-673-662, (1979).

McInnes, Colin R., *Solar Sailing: Technology, Dynamics and Mission Applications,* Praxis Publishing, Chichester, UK, 1999.

Manne, Alan S. and Richels, Richard G., "CO^2 Emission Limits: An Economic Cost Analysis for the USA" in *The Energy Journal,* edited by Leonard Waverman, Washington, D.C., 11(2), April 1990, pp. 51-74.

Mallove, E., and Matloff, G.L., *The Starflight Handbook,* Wiley (1989).

Matloff, Gregory, *Deep-Space Probes,* 2nd ed., Springer-Praxis (2005).

Matloff, G. L., Johnson, L., and Bangs, C, *Living Off the Land in Space,* Springer-Praxis (2007).

Roy, K.I., "Solar Sails: An Answer to Global Warming?", *CP552, Space Technology and Applications International Forum-2001,* edited by M.S. El-Genk (2001).

Roy, Ken and Kennedy, Robert G., "Mirrors & Smoke: Ameliorating Climate Change With Giant Solar Sails", *Whole Earth Review,* edited by Bruce Sterling, Summer 2001, p.70.

Stark, John P.W., "Celestial Mechanics" in *Spacecraft Systems Engineering,* edited by Peter Fortescue and John Stark, John Wiley and Sons, Chichester, 1991, pp. 59-81.

Vulpetti, G., Johnson, L., and Matloff, G. L., *Solar Sails; A Novel Approach to Interplanetary Travel,* Springer-Praxis (2008).

Chapter 17
Settling the Solar System

The Multi-Planet Species

Today, the biosphere is bounded by Earth's
    atmosphere.
But this may not always be so.
Gaia may evolve and expand.
Using technology, life can move beyond
    Earth
    to inhabit lunar lava tubes
    and domed cities on Mars
    and habitats constructed from asteroid
    material.
Energy is abundant in space,
water and oxygen exist in many places.
Someday, we will be a multi-planet species
    citizens of the solar system
    and the wide cosmos beyond.

G. Matloff et al., *Harvesting Space for a Greener Earth*,
DOI 10.1007/978-1-4614-9426-3_17, © Springer Science+Business Media New York 2014

"Water, water everywhere,
And all the boards did shrink;
Water, water everywhere
Nor any drop to drink"
—From the poem "The Rhyme of the Ancient Mariner" by
Samuel Taylor Coleridge

The thought is not new, but it is profound, and it can help guide humanity as we deal with the myriad environmental challenges facing us at the beginning of the twenty-first century. We all inhabit one giant spaceship on a voyage together through the dark emptiness of space. We call it "Spaceship Earth."

Clearly visible from the hotels in Orlando, Florida, is Disney's Spaceship Earth at the Epcot Center (Fig. 17.1). When visiting the dome, tourists experience the history of human communication, from the dawn of humankind to today. For many, seeing the dome reminds them of the interconnection we have with each other and with our home planet—a more far-reaching outcome than is intended.

As we learn to be better stewards of Earth and thereby protect it for habitation and use by future generations, we will also be learning how to better extend the human presence beyond Earth and into the nearby Solar System. The reverse process is also true. Many of the technologies being developed to support human life beyond Earth may have application here at home as well. This chapter examines the essential parts of a self-sustaining outer space habitat and discusses how this relates to similar processes on Spaceship Earth.

The essential elements of a crewed space mission include Earth-to-orbit transportation, transportation in space and away from Earth, energy, food, water, waste disposal or recycling, and the commodities of everyday life. We will not dwell on the transportation systems required for space exploration. For more complete background on both Earth-to-orbit and in-space transportation systems, see the authors' previous books, *Living Off the Land in Space* (New York: Springer-Praxis, 2007) and *Solar Sailing: A Novel Approach to Interplanetary Travel* (New York: Springer-Praxis, 2008). We will discuss the other issues cited above, as all of them are linked to terrestrial analogs with which they have much common ground.

Water

As we explore space, it is vital that we find water. We humans need it not only for drinking but also for bathing, cooking, and for use in our rockets. The Apollo astronauts took with them enough water for the round trip to and from the Moon, with some in reserve. The International Space Station recycles some of its water, but must still be regularly resupplied from Earth. Any long-term settlement will need a supply of water and a way to recycle it.

Fig. 17.1 Spaceship Earth at the Disney World Epcot theme park reminds visitors of the giant spacecraft we all inhabit (courtesy of Katie Rommel-Esham)

Here are some water facts:

- Approximately 70.8 % of Earth's surface is covered by water—roughly 1,260,000,000,000,000,000,000 L.
- 98 % of the water on the planet is in the oceans.
- 1.6 % is in the polar ice caps and glaciers.
- 0.36 % is found underground (in aquifers and wells).
- 0.036 % is found in lakes and rivers.
- By weight, our bodies are approximately 62 % water.
- You cannot live more than a few days without water, depending on the temperature, your activity level, and your general state of health.
- Water can be separated into its constituents, hydrogen and oxygen, and used as rocket fuel.

Fortunately, nature provides us with water throughout the Solar System. Early Earth is thought to have received its water from frequent comet impacts. Other planets and moons experienced similar impacts and therefore should have water as well. We now know that there is water on Mars, locked mostly in subsurface ice and in the planet's polar ice caps. The Moon may have water in the perpetually shaded craters at its south pole. Based on recent observations, the main belt asteroid Ceres may be 25 % water. The many comets that periodically visit the inner Solar System are composed mostly of water ice.

Just because there is water at potential settlement destinations does not mean that we can use it profligately. As discussed in other chapters, water is a valuable resource and one that must be used wisely with an eye toward recycling. There simply is not enough of it out there, nor will what is there be easily accessed.

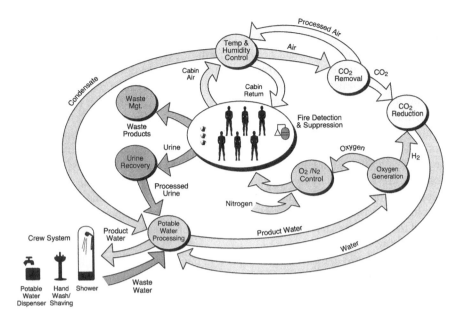

Fig. 17.2 The space station's Environmental Control and Life Support System regenerates both air and water for the crew (image courtesy of NASA)

Nothing accomplished in space is easy, and every attempt must be made to use and reuse the resources already "in hand."

NASA has been working on the water problem for years and is implementing a water recycling system on the International Space Station (ISS). Otherwise, NASA and its partners would have to launch about 18,000 kg of water annually just to keep the station functioning. The water recycling system, technically known as the Environmental Control and Life Support Systems Water Recycling System (ECLSS-WRS), reclaims and prepares for reuse water from the crew's urine, showers, and that which is exhaled into the air, as shown in Fig. 17.2. (When on the ISS, don't ask where that cool drink of water came from!)

Water on Earth is recycled, although we do not usually think of it in that way. On Earth, water passes through our bodies and is put back into the environment. Microbes in the soil break down our wastewater and convert the waste part of it into nutrients that plants can use and absorb to grow. These microbes and the soil act as filters, purifying the water for later consumption by someone or something else. Some of our wastewater evaporates and is filtered by atmospheric processes before it falls back to the ground as rain or snow.

Scientists studied nature to design the systems our space explorers will use to recycle water. The techniques developed for space are now being used to help improve terrestrial water treatment plants' efficiency and efficacy.

New ways to extract water from inhospitable places are being developed and tested in anticipation that they might be used by future explorers who truly want to learn how to live off the land and not be dependent upon the supply chain from

Earth. Water ice has long been thought to exist within craters at the lunar south pole. It may not be simply sitting there in convenient cubes ready to haul back to a human outpost; rather, it might be in the form of permafrost beneath the first few centimeters and mixed thoroughly with the rock that makes up the lunar regolith. The question then becomes, how do you get the water from the rock?

The answer may be as simple as your kitchen's microwave oven. Recent experiments at the NASA Marshall Space Flight Center show that microwave heating of the regolith might release as much as 99 % of the any trapped water contained within it. The NASA team set up an experiment using simulated lunar regolith and had some interesting and promising results. Basically, the team 'cooked' the soil using a converted microwave oven. The energy of the microwaves penetrated the soil and heated it from the inside out. The team found that when the regolith heated from -150 °C to about $-50°$, any water trapped in the soil was forced to the surface and escaped as vapor. The vapor was collected and condensed into liquid water.

Can this be used to extract water trapped in the desserts of Earth? If so, can it be done affordably? And where would the energy come from to power the microwave devices? These questions will have to be answered before it can be seriously attempted.

Air

Astronauts need air to breathe when they are in space. Earth's envelope of breathable air extends only a few kilometers upward and then thins to almost nothing at 100 km (Fig. 17.3). Think about that. The air upon which all life on Earth depends essentially ends only 100 km above our heads. Furthermore, half of all the air and nearly 90% of all the moisture within the atmosphere is trapped within the first 6 km of that 100.

When walking, the average man breathes approximately 30 L of air per minute. The average woman requires 20 L per minute. For both sexes, the amount required increases dramatically when exercising. Here are some air facts:

- The air we breathe is approximately 78 % nitrogen and 21 % oxygen. Carbon dioxide (CO_2) is present in trace amounts.
- 75 % of Earth's atmospheric mass is less than 11 km above its surface.
- Trees make their own food from the CO_2 in the air and release oxygen for us to breathe.
- One acre of forest can produce enough oxygen to sustain 18 people each day.
- You cannot live more than just a few minutes without air.

Where will future space explorers get their air? The air we breathe here on Earth is recycled. The oxygen in the air is the part that we consume and turn into CO_2 during respiration. We breathe in air, mostly composed of nitrogen and oxygen, and we breathe out air composed of nitrogen and CO_2. By removing the carbon dioxide

Fig. 17.3 The full Moon was photographed above the tenuous atmospheric layer of Earth by astronauts on the International Space Station in 2011 (image courtesy of NASA)

and replenishing the oxygen, we can maintain a breathable atmosphere. The problem for space explorers, therefore, is twofold: getting rid of the CO_2 and replacing it with oxygen.

Onboard the ISS and in most long-duration human space flights, CO_2 is removed from the air and dumped overboard. The oxygen is replenished with resupply from Earth. This clearly will not work for long-duration missions in deep space. Where, then, will explorers get their oxygen? The answer to this question depends on where the explorers are going.

As mentioned above, the Moon may contain water in the shadowed craters near its solar pole. Without sunlight, the remains of long-ago comet strikes may lie dormant deep within these craters, providing a large supply of ice for future explorers. Water and ice are composed of hydrogen and oxygen. These chemicals may be separated when electrical current flows through the water in a process called electrolysis. Electrolysis is not new. Since its discovery in 1800, electrolysis has been used in many industrial processes and may be used in the future to mass-produce hydrogen for Earth-based fuel cells (Fig. 17.4).

Anywhere our explorers and settlers encounter water, they can use the energy of sunlight to produce electricity and, through electrolysis, oxygen. But what about places were there is little or no water? When in deep space, will explorers be able to find other sources of oxygen?

Again, our nearest neighbor, the Moon, has additional resources that might be used to produce oxygen. Many lunar rocks contain the mineral ilmenite, which is made of iron titanium oxide. Thanks to the Apollo missions, we know that ilmenite is plentiful on the Moon. To release the oxygen trapped within the rocks, it is in

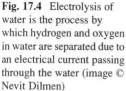

Fig. 17.4 Electrolysis of water is the process by which hydrogen and oxygen in water are separated due to an electrical current passing through the water (image © Nevit Dilmen)

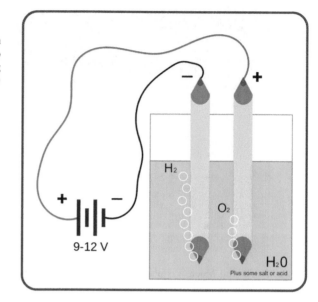

principle as easy as adding hydrogen and heat (from sunlight). The technologies for accomplishing this are in hand and could fairly easily be adapted for use in space. For example, an inflatable solar concentrator (think of it as a balloon shaped like a magnifying glass) could concentrate sunlight on a sample in a hydrogen-filled pressure vessel, heating the rock to greater than 1,000 °C. The process would produce iron, titanium oxide, and water. The water could then be converted to oxygen and hydrogen.

An innovative way to extract oxygen from the rock using ionic liquids, which work at much lower temperatures, is also being considered. Ionic liquids are composed of oppositely charged ions (atoms with an electrical charge)—salts may be ionic liquids. They typically have low melting temperatures (~100 °C) and can be regenerated after use, making them ideal for use in situations where resupply of materials is difficult.

First, the regolith would be ground to a fine, sandy consistency and then introduced to an ionic liquid at a temperature of approximately 200° C. The mixture sits for about a day, during which time a hydrogen atom from the ionic liquid bonds to the oxygen atoms of the rock to form water vapor. The vapor is extracted and the oxygen is removed by electrolysis. The other part of the water molecule, hydrogen, is pumped back into the mixture to regenerate the ionic liquid for use in the next batch of rock to be processed.

Earth is very efficient at recycling CO_2, though perhaps not efficient enough. The rate at which human industrial civilization is pumping CO_2 into the atmosphere appears to exceed the rate at which the currently established natural processes can remove it. Application of proposed carbon sequestration and reuse technologies could benefit both Earth and our deep space explorers. For example, captured CO_2

from a power plant or from astronaut respiration could be used to produce methanol. Methanol, an alcohol, could be used as a fuel for transportation or electricity-producing fuel cells. Today, methanol is synthesized directly from fossil fuels, but tomorrow it could be made from the waste produced by burning fossil fuels or from the very air we exhale.

Power

The key to space settlement is power. Power is required for the transportation systems that will get our settlers into space and to their final destinations. Power will be required to keep them from getting too warm or too cold. It will be needed to run their machines, produce and recycle water, replenish the oxygen in the air they will breathe, and power their scientific instruments. It will be needed in every facet of space exploration and settlement and cannot be taken for granted, as early Earth-bound explorers could do, knowing as they did that the Sun would rise the next day to provide them with light and heat. In space, taking advantage of the power from space will not be as easy as merely basking in the sunlight from one day to the next.

Many of the technologies our space settlers will use to generate power are closely analogous with their terrestrial counterparts. For settlements near the Sun, which is all we can expect for at least the next 100 years or so, solar power should be the primary option. On a small scale, our explorers can extend arrays of solar cells outward from their habitats to gather as much sunlight as possible for direct conversion to electricity. Near Earth, the energy density of sunlight is 1,368 watts per square meter. With no cloudy days, space settlers should be able to convert this perpetual source of energy into electricity with close to 40 % efficiency.

Solar cells are primarily made from silicon, which is abundant on both Earth and the Moon. On the Moon, solar cells might therefore someday be made from the lunar soil in much the same way they are made from sand here on Earth. Large solar arrays can be deployed on the surface of the Moon to generate most of the power needed for a human settlement. An artist's conception of a solar array "farm" near a lunar settlement is shown in Fig. 17.5.

Unfortunately, the length of time the sunlight is available depends on the location, and some locations are sunnier than others. In low Earth orbit, where the first space industrial parks may someday be constructed, orbiting facilities will experience darkness 40–50 % of the time as they enter and travel through Earth's shadow. (Solar cells do not produce any power when there is no sunlight.) To keep the power on during these frequent eclipses, technologists are developing better batteries that can store and release large amounts of power repeatedly with high efficiency.

One of the most promising of these new batteries is called a flywheel, which is a mechanical device that stores energy in the rotation, or spin, of an internal wheel. When power is put into a flywheel, the rate of spin increases. When power is

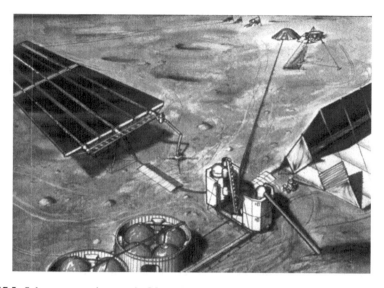

Fig. 17.5 Solar power stations on the Moon may provide much of the power required for future human settlements (courtesy of the Space Studies Institute)

removed, the spin rate decreases. Unlike virtually all conventional batteries, including fuel cells, no chemical processes are involved. Flywheels are strictly mechanical, and they can theoretically store very large amounts of energy with frequent charge and discharge cycles, suffering little or no loss of efficiency as they age.

The power system of an Earth-orbiting industrial park would have to generate roughly twice as much power as it needs during the sunlit portion of its orbit so that it can store 50 % of the power generated by spinning up its flywheel batteries. When the park enters Earth's shadow and the power from its solar cells drops to zero, the energy stored in the flywheels is tapped to keep the station running until sunlight returns. This spin-up and spin-down cycle is repeated every time the station completes an orbit.

Flywheels could also be used on the Moon, where the day/night problem is much more severe. The Moon's rotation is much slower than Earth's. As a result, the length of a lunar day is much longer than its terrestrial counterpart. Instead of 24 h, the lunar day is 28 Earth days long, with almost any given lunar location experiencing approximately 50 % of that time in darkness. When you depend on the Sun for power, a 14-day "night" is a long night indeed. The flywheel energy storage system required for the Moon will be very large.

The situation faced by space habitats needing supplemental power when they are not able to generate it on their own is analogous to that faced by terrestrial power stations during their peak loading periods. Consider, for example, the power demand placed on utilities in the American south during the summer months. The peak demand for power occurs in the mid-afternoon, when the Sun is high and office buildings, stores, and factories are using their air conditioners. This is also the time that some workers are starting to go home and are turning on their home air conditioners. The demand for electrical power during this time often outstrips the ability of the utility to generate it. This results in the importing of expensive power from outside the local grid, thereby raising costs. Alternatively, if no supplemental power is available, voluntary or involuntary brown-outs might occur.

If large flywheels could be installed to store power generated during times of lower power demand (for instance, at night), and release it during these daytime peaks, then the need for expensive non-utility-generated power could be reduced and the occasional brown-out avoided. On-demand power and efficient power storage are required for both space and Earth.

Finally, the space solar power stations described in Chap. 12 could beam energy to Earth-orbiting or lunar habitats as required. Once a space-based power infrastructure is in place, the uses for it will multiply.

Radiation Protection

Space is a harsh and deadly radiation environment from which humans need protection—here on Earth and in our space settlements. The Sun produces far more radiation than we can see using our eyes. The fusion-driven furnace within the Sun is producing X-rays, gamma rays, and highly energetic protons, electrons, and alpha particles—all of which are permeating the near-space environment in such quantities as to pose a threat to human health from near the Sun to Earth, Mars, and beyond. Without some form of shielding (from this radiation), life cannot survive.

Earth does an excellent job of protecting us from most of this deadly radiation. The same magnetic field that makes our compass needles point toward the North Pole acts as a *Star Trek*-like deflector screen, diverting and trapping much of the deadly radiation pummeling our planet on a daily basis. We can even observe some of this trapped radiation as it enters our atmosphere and interacts with it near the North and South Poles; we call it the aurora borealis. The aurora is nothing more than high-energy charged particles traveling along Earth's magnetic field careening into the dense atmosphere, ionizing it at high altitudes, and causing atmospheric oxygen to glow.

Although Earth's magnetic field will not stop all of the electromagnetic radiation from the Sun, our thick atmosphere absorbs or attenuates much of what is left. Only visible light and some of the ultraviolet can get through the kilometers of air that separate us from the vacuum of space. Fortunately, the chemical composition of the atmosphere allows it to absorb much of the ultraviolet light through its interaction with ozone. Ozone is a naturally occurring form of oxygen produced in the upper atmosphere. In the 1970s, satellite data showed that Earth's ozone layer was depleting, producing potentially dangerous ozone "holes" over some parts of the globe. These holes allowed more ultraviolet light to reach Earth's surface, potentially causing increased incidences of cancer. Many scientists believed that certain chemicals commonly produced and used in industrial and commercial processes caused the depletion of the ozone layer. The body of evidence supporting this view was enough to convince politicians that the use of these ozone-depleting chemicals should be reduced. Laws to reduce their use were enacted, and the slow recovery of the ozone layer is thought to be occurring today.

Space explorers will not have the benefit of kilometers of atmosphere, magnetic field deflector screens, and ozone layers to protect them. Instead, they will have to rely on artificial shielding and advanced warning of impending solar storms, which are increased periods of high radiation, so that they can take shelter and not experience potentially lethal exposure to the solar radiation. Interestingly enough, the same solar storm warning system that space settlers will require is in use today, providing spacecraft operators and utility companies with advanced warning of solar storms so that they do not experience catastrophic failures themselves.

Before one can understand how a solar storm can cause the lights to go out on Earth, some aspects of physics that are out of the realm of most people's everyday experiences must be explained. The burst of radiation that is a solar storm is composed, in large part, of charged particles: protons, alpha particles, and electrons. These moving charged particles generate a strong and massive magnetic field. When the storm encounters Earth, it tends to push Earth's magnetic field inward toward the surface, lowering the altitude of the magnetic field lines dramatically.

When charged particles move through a magnetic field, they experience forces acting upon them. The reverse is also true. A moving magnetic field induces a current flow in a wire. A current is nothing more than a flow of electrical charges through a wire or some other conducting medium. Electrical utility wires (particularly those hanging from telephone poles at northern latitudes) feel the effect of the solar storm as Earth's magnetic field is compressed toward Earth, changing in intensity with time. This changing magnetic field induces current flow in the wires and voltage differences at the various grounding points within the power grid, creating spurious currents that knock out transformers and otherwise disrupt or shut down the transmission of electrical power.

This is a real effect, and it has happened. On March 13, 1989, a solar storm sent the Hydro-Quebec, Canada, power grid, which serves more than 6 million people, into a blackout. These storms are such a threat that utilities monitor space weather conditions so that they can have repair crews available to fix the inevitable damage to their infrastructure when such storms are predicted to arrive.

To provide at least some warning of impending solar storms, the National Oceanic and Atmospheric Administration (NOAA) and NASA placed the Advanced Composition Explorer (ACE) spacecraft at one of the Earth-Sun Lagrange points. A spacecraft placed here will likely remain unless some outside force acts upon it. The regions are not 100 % gravity or disturbance free, so some spacecraft propulsion is required to remain within them. The fuel required, however, is much less than would be needed should these regions not exist.

The ACE spacecraft detects a solar storm when radiation from an associated event on the Sun strikes its detectors; the spacecraft then sends a radio signal back to Earth, also traveling at the speed of light, telling satellite operators and even Earth-bound electrical power utilities that a storm is coming and to get ready. The light from the Sun, and subsequently the radio transmission signaling the impending storm, reach Earth about one hour before the ionizing radiation because light travels faster in the vacuum of space than do the charged particles originating from the Sun. This same sort of spacecraft will send a warning to our space settlers, urging them to take shelter.

Space settlers must also contend with another form of radiation: galactic cosmic rays. These electrically charged particles are accelerated to near-light-speed velocities by cosmic electromagnetic fields. Because only about two dozen humans have thus far voyaged beyond the protective confines of Earth's magnetic field, we have little data regarding the safe threshold for human exposure to these particles. The best methods of protecting against galactic cosmic rays are either to equip the space settlement with a massive, thick shield of rock or soil gathered from celestial bodies, or to generate an artificial magnetic field. As humans travel once again to the Moon (and other Solar-System bodies), space mission planners will learn a great deal more about this radiation source.

As we learn how to explore space, we will learn more about how to improve life on Earth. From taking the technologies developed to purify water on the International Space Station and adapting them for use in terrestrial water purification systems to the new technologies for capturing and sequestering carbon dioxide, the tools we will need for space development are increasingly becoming the same tools we need here on Earth.

Further Reading

Appendix 1 contains a NASA Fact Sheet describing the International Space Station's Environmental Control and Life Support System.

For a detailed discussion of our vulnerability to space weather, please refer to Mark Moldwin's excellent introductory book on the subject aptly titled, *An Introduction to Space Weather*. Up-to-the-minute space weather information can be found at the NOAA Space Weather Prediction Center online at http://www.swpc.noaa.gov/.

Chapter 18
An Optimistic Future

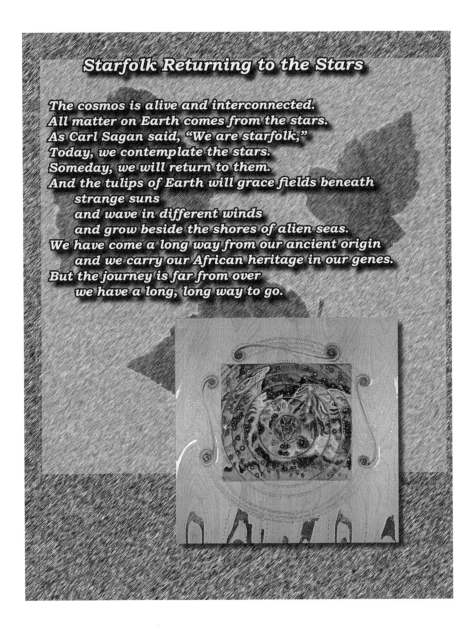

Starfolk Returning to the Stars

The cosmos is alive and interconnected.
All matter on Earth comes from the stars.
As Carl Sagan said, "We are starfolk,"
Today, we contemplate the stars.
Someday, we will return to them.
And the tulips of Earth will grace fields beneath
    strange suns
    and wave in different winds
    and grow beside the shores of alien seas.
We have come a long way from our ancient origin
    and we carry our African heritage in our genes.
But the journey is far from over
    we have a long, long way to go.

G. Matloff et al., *Harvesting Space for a Greener Earth*,
DOI 10.1007/978-1-4614-9426-3_18, © Springer Science+Business Media New York 2014

"Oh yesterday the cutting edge drank thirstily and deep,
The upland outlaws ringed us and herded us like sheep,
They drove us from the stricken field and bayed us into keep;
But tomorrow,
By the living God, we'll try the game again!"
—From the poem "Tomorrow" by John Masefield,

Gloom and doom is easy to sell. Simply watching the evening news fills one with a sense of what is bad in the world: murder, child abuse, increases in the price of gasoline, water rationing due to drought, tension among nation states arising over a religious or economic disagreement, and so on. Listening to the prognostication of the coming age of scarcity and environmental disaster makes for great drama and feeds what seems like an increasing sense of global pessimism.

Yet, while the gloom and doom sells newspapers and movies, progress marches onward. More people today are free than at any other time in history, the standard of living for the average person on Earth has steadily increased, and many diseases that once ravaged the human population no longer exist. The human condition is undoubtedly improving, and we should revel in it.

However, the warning signs for potential danger are real. Leaving aside the hotly debated question of the degree to which we humans are contributing to climate change and global warming, we posit that all humans should be concerned about Earth's environment and its overall health. It is our one and only home for the foreseeable future, and we should do all that we can to protect it and preserve it for future generations. We should personally reduce our trash, recycle what we can, and use sparingly what we cannot. We should recycle, reuse, and efficiently consume only as much as we need, and we argue that this can be done without causing lowered living standards if we make our investments wisely and choose a path that leads to increased prosperity and living conditions for all humanity. It is here that space advocates can work with environmentalists to make that prosperous and Earth-friendly future a reality.

By working to decrease the cost of launching payloads into space, we will enable that constellation of space solar power satellites to be built around Earth, beaming downward nearly infinite energy to provide power for a prosperous humanity. With increased energy availability has historically come an increase in the standard of living. Power provided from space will allow the eventual closure of oil- and coal-burning power plants, reducing the level of atmospheric pollution and global CO_2 emissions. At night, people around the world will have a visible sign of progress gleaming in a ring around the world as they look skyward. Reflecting sunlight from their massive solar arrays, the solar electric power stations will serve as an inspiration for the next generation as they envision where we will next go in our exploration and utilization of space.

Will the cost of space access decrease? Absolutely. Studies have repeatedly shown that the most reliable way to reduce the cost of launching satellites into space is to increase the flight rate. Companies such as Space X, Blue Origin, and XCor Aerospace are not just talking about developing new rockets; they are building and launching them. The result? Launch prices are already coming down, and the currently entrenched industry leaders will either become more price competitive or see their customer base dwindle.

Residents near the world's space launch sites will see the flow of material rocket skyward toward a destination between Earth and the Sun as the world's sunscreen is erected there. Stopping a mere 0.25 % of the light reaching Earth will reduce the overall energy input to the planet, providing an offset to the excess heat trapped near the surface from profligate greenhouse gas emissions in the twentieth and early twenty-first centuries. Earthbound residents will have reduced their net greenhouse gas emissions to near zero from recycling, reuse, and space solar power. But to actually reduce the levels of atmospheric CO_2 to preindustrial levels will be well beyond our reach for quite some time. To buy us the needed time, the Sun shield will not only halt increases in global temperatures but also reduce them to levels experienced prior to the industrial revolution.

Privately funded space ventures will be visiting nearby asteroids to mine their resources and bring them back home for use on Earth. Companies such as Planetary Resources, an ambitious venture with the backing of well-known and pedigreed space advocates such as Eric Anderson and Peter Diamandes (of X-Prize fame) and the financial backing of Larry Page and Ross Perot, Jr., have the potential to revolutionize the global supply chain of raw materials. They are not alone in trying to mine the near-space asteroids. The Deep Space Industries mission statement says, "Deep Space Industries believes the human race is ready to begin harvesting the resources of space both for their use in space and to increase the wealth and prosperity of the people of planet Earth."

The authors of this book wish them both success, for if they are, the entire planet will benefit. Instead of strip mining entire mountains, re-greening of previously mined mountains can begin. The need for raw materials will not decrease as prosperity increases; rather, it, too, will increase. A steady supply of raw materials will be shipped toward space or lunar-based processing and manufacturing facilities to meet the demand of Earth-bound consumers. These consumers will use the products made from off-world resources and then, with ever-increasing efficiency and simplicity, recycle them for reuse.

Yet other missions will be launched to potentially Earth-threatening asteroids to deflect them from collision with Earth to more benign orbits for either mining or never to be seen again. The specter of global catastrophe from asteroid collision will be eradicated from the list of threats to the human race. Here, too, private space ventures are working to make a difference. The B612 Foundation is funding the development of an infrared space telescope to discover and track near Earth asteroids (NEA's) that may someday threaten the planet. Such a telescope will be vital to locate those last few asteroids that we might not otherwise detect due to them coming toward us from the direction of the Sun—otherwise lost in the glare.

Robots in Earth-orbiting factories or those on the Moon will accomplish more and more of our manufacturing. Industrial sites that once marred the landscape can be taken apart and the land reclaimed. At first only the most environmentally hazardous facilities will be located off-planet, with their often-unavoidable toxic by-products forever banished to the vastness of empty space or on a collision course with the Sun rather than with a living, breathing Earth. As accessing, living in, and working in space become more commonplace, more industry will be moved there in order to be closer to their primary source of raw materials (asteroids) and power (the Sun). Eventually, more and more people on Earth will not have to live and love in the shadow of smokestacks and toxic dumps.

All the while, a small fleet of spacecraft will be circling Earth, monitoring the environment, and providing a continuing assessment of our progress toward reclaiming Earth and returning it to be what it should be for humanity—a place to live, a paradise regained.

Afterword: Why Space Advocates and Environmentalists Should Work Together

Aerospace is not usually considered verdant. Jet aircraft emit tons of carbon dioxide into the atmosphere; some of the companies that make them are defense contractors, and nobody considers weapons of war, necessary or not, to be green. These companies are in business to make money for their shareholders and will only be green if it somehow benefits their bottom line or if law or mandate requires them to be so. This reality, and there are exceptions, also taints those involved in space exploration and development. But painting the entire canvas with the same brush is not only grossly unfair, it is also incorrect.

Many people who devote their lives and careers to the exploration of space are not in it for the money. Sure, they need to earn a living like everyone else. But that is not why the authors of this book chose their careers. Greg Matloff is an astronomer and astronomy professor. He lives, breathes, studies, and teaches about the heavens. He does this, in part, because studying the universe tells us something about ourselves. As we have explored the universe around us, we have learned how precious our planet Earth truly is.

C Bangs explores archetypes of Earth and depicts cosmological elements from an ecological and feminist perspective. One premise of her work is that we are part of Earth and all the elements of our bodies at one time were within a star. We contain both systems within us. Her art in a wide variety of media is informed by mythology and the hope for human evolution. Space is not only about individual heroics or national pride. Ultimately, it deals with the expansion of consciousness and the survival of the human species and other terrestrial life forms.

For example, by studying the climate of Mars, we have something to which we can compare Earth's climate. This provides us with another data point, and that is vitally important because it is difficult, if not impossible, to explain anything generally with only one example. An analogy would be trying to draw a line using only one point. You are free to draw such a line in any direction within a 360° angle. Making conclusions about any specific line that one draws as being "correct" is therefore impossible. With two points, you may draw a line. It may still not accurately describe the system, but it will likely be closer to correct than any

G. Matloff et al., *Harvesting Space for a Greener Earth*,
DOI 10.1007/978-1-4614-9426-3, © Springer Science+Business Media New York 2014

other of the infinite set of lines that could have otherwise been drawn through a single point.

Les Johnson is a NASA physicist and manager. He dreamed of working for NASA since he was a 7-year-old boy in Ashland, Kentucky, and watched on television as Neil Armstrong set foot on the Moon. He knew then that he wanted to be part of that great adventure and that he had to become a scientist in order to do so.

The three of us are not unique, nor are we in the minority among those who work in the field of space exploration and science. Most of our colleagues with whom we have discussed career motivations ("Why did you study science and become an astronomer/physicist/engineer?") share our fundamental love of knowledge and passion to explore, develop, and use space for the betterment of humankind. And this passion is not limited to scientists and engineers.

Businessman Robert Bigelow is using the fortune he earned in business to foster space development. He is the founder and owner of Bigelow Aerospace near Las Vegas, Nevada, which is building what may be the world's first orbital hotel. Bigelow has already flown subscale prototypes into Earth orbit and has plans to loft full-scale pressurized modules within the near future.

Another businessman, Elon Musk, is using his dot.com wealth to develop a new generation of inexpensive rockets to carry machines and eventually people into space at a lower cost than is currently possible.

Environmentalists share the vision of protecting and preserving life on Earth. By working together we stand a much better chance of being able to make this vision a reality.

As we gaze skyward on a starry night, we are struck by how immense and seemingly lifeless the universe appears to be. If we are not alone among the stars, then we are among a very few civilizations likely separated by an abyss that will be impossible to cross. Hence the imperative that we preserve that which makes us unique. We are part of a living planet. Exploring and developing space will help us make this a better, greener world.

Brooklyn, NY, USA   Greg Matloff
Brooklyn, NY, USA   C Bangs
Madison, Al, USA   Les Johnson

Appendix 1: The Current NASA Approach to Human Life Support in Space

To Dance Within a Diamond Sky!

Imagine being weightless
You float through the cabin
Drift effortlessly through the air
Every move becomes a dance
You are like a bird, or a bee.
What exhilaration!
What freedom!
Today, only a few hundred people have
    danced like this.
But someday it may be the birthright of
    millions.

G. Matloff et al., *Harvesting Space for a Greener Earth*,
DOI 10.1007/978-1-4614-9426-3, © Springer Science+Business Media New York 2014

National Aeronautics and Space Administration

International Space Station
Environmental Control and Life Support System

NASA's Marshall Space Flight Center in Huntsville, Ala., is responsible for the design, construction and testing of regenerative life support hardware for the International Space Station, as well as providing technical support for other systems that will provide the crew with a comfortable environment and minimize the resupply burden.

The Environmental Control and Life Support System (ECLSS) for the Space Station performs several functions:

- Provides oxygen for metabolic consumption;
- Provides potable water for consumption, food preparation and hygiene uses;
- Removes carbon dioxide from the cabin air;
- Filters particulates and microorganisms from the cabin air;
- Removes volatile organic trace gases from the cabin air;
- Monitors and controls cabin air partial pressures of nitrogen, oxygen, carbon dioxide, methane, hydrogen and water vapor;
- Maintains total cabin pressure;
- Maintains cabin temperature and humidity levels;
- Distributes cabin air between connected modules.

Background

Earth's natural life support system provides the air we breathe, the water we drink and other conditions that support life. For people to live in space, however, these functions must be done by artificial means.

The life support systems on the Mercury, Gemini and Apollo spacecraft in the 1960s were designed to be used once and discarded. Oxygen for breathing was provided from high pressure or cryogenic storage tanks. Carbon dioxide was

removed from the air by lithium hydroxide in replaceable canisters. Contaminants in the air were removed by replaceable filters and activated charcoal integrated with the lithium hydroxide canisters. Water for the Mercury and Gemini missions was stored in tanks, while fuel cells on the Apollo spacecraft produced electricity and provided water as a byproduct. Urine and waste-water were collected and stored or vented overboard.

The Space Shuttle is a reusable vehicle, unlike those earlier spacecraft, and its life support system incorporates some advances. But it still relies heavily on the use of consumables, limiting the time it can stay in space.

The International Space Station includes further advances in life support technology and relies on a combination of expendable and limited regenerative life support technologies located in the U.S. Destiny lab module and the Russian Zvezda service module. Advances include the development of regenerable methods for supplying

NASAfacts

www.nasa.gov

oxygen (by electrolysis of water) and water (by recovering potable water from wastewater).

These advances will help to reduce the cost of operating the Space Station because it is expensive to continue launching fresh supplies of air, water and expendable life support equipment to the Station and returning used equipment to Earth.

Providing Clean Water and Air

The Space Station Environmental Control and Life Support System includes two key components – the Water Recovery System (WRS) and the Oxygen Generation System (OGS). The systems have been jointly designed and tested by the Marshall Center and Hamilton Sundstrand Space Systems International in Windsor Locks, Conn. They are packaged into three refrigerator-sized racks that will be located in the U.S. Lab of the Station.

The Water Recovery System provides clean water by reclaiming wastewater, including water from crewmember urine; cabin humidity condensate; and Extra Vehicular Activity (EVA) wastes. The recovered water must meet stringent purity standards before it can be used to support crew, EVA, and payload activities.

The Water Recovery System is designed to recycle crewmember urine and wastewater for reuse as clean water. By doing so, the system reduces the net mass of water and consumables that would need to be launched from Earth to support six crewmembers by 15,000 pounds (6800 kg) per year.

MSFC ECLSS Test Facility

Water Processing Flight Experiment

Space Station Water Processor Test Area

Urine Processor Flight Experiment

The Water Recovery System consists of a Urine Processor Assembly (UPA) and a Water Processor Assembly (WPA). A low pressure vacuum distillation process is used to recover water from urine. The entire process occurs within a rotating distillation assembly that compensates for the absence of gravity and therefore aids in the separation of liquids and gases in space. Product water from the Urine Processor is combined with all other wastewaters and delivered to the Water Processor for treatment. The Water Processor removes free gas and solid materials such as hair and lint, before the water goes through a series of multifiltration beds for further purification. Any remaining organic contaminants and microorganisms are removed by a high-temperature catalytic reactor assembly. The purity of product water is checked by electrical conductivity sensors (the conductivity of water is increased by the presence of typical contaminants). Unacceptable water is reprocessed, and clean water is sent to a storage tank, ready for use by the crew.

The Oxygen Generation System produces oxygen for breathing air for the crew and laboratory animals, as well as for replacement of oxygen lost due to experiment use, airlock depressurization, module leakage, and carbon dioxide venting. The system consists mainly of the Oxygen Generation Assembly (OGA) and a Power Supply Module.

The heart of the Oxygen Generation Assembly is the cell stack, which electrolyzes, or breaks apart, water provided by the Water Recovery System, yielding oxygen and hydrogen as byproducts. The oxygen is delivered to the cabin atmosphere while the hydrogen is vented overboard. The Power Supply Module provides the power needed by the Oxygen Generation Assembly to electrolyze the water.

The Oxygen Generation System is designed to generate oxygen at a selectable rate and is capable of operating both continuously and cyclically. It provides from 5 to 20 pounds (2.3 to 9 kg) of oxygen per day during continuous operation and a normal rate of 12 pounds (5.4 kg) of oxygen per day during cyclic operation.

The Oxygen Generation System will accommodate the testing of an experimental Carbon Dioxide Reduction Assembly (CReA). Once deployed, the reduction assembly will cause hydrogen produced by the Oxygen Generation Assembly to react with carbon dioxide removed from the cabin atmosphere to produce water and methane. This water will be available for processing and reuse, thereby further reducing the amount of water to be resupplied to the Space Station from the ground.

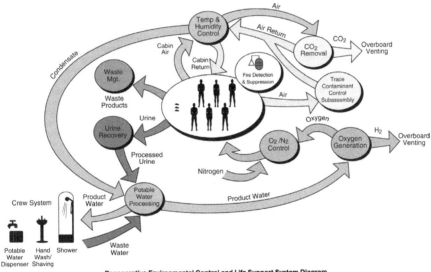

Regenerative Environmental Control and Life Support System Diagram

Prototype ECLSS Racks

Ground testing before flight

Before the Environmental Control Life Support System hardware is launched to the International Space Station and put into operation, it will undergo extensive testing on the ground and in space.

The Marshall Center maintains facilities for testing and evaluating life support technologies. These facilities also allow engineers on the ground to troubleshoot any problems encountered in space. Testing includes life cycle tests to determine maintenance and change-out requirements, operational tests, such as those using volunteers to generate waste-water for the Water Recovery System, and integration tests of flight hardware for the Space Station.

Space-proven hardware

Portions of the Environmental Control and Life Support System hardware have been flight-tested before being installed aboard the Station. The Volatile Removal Assembly, for example, flew aboard STS-89 in January 1998, to demonstrate the microgravity performance of the Water Processor Assembly's catalytic reactor.

The Marshall Center also was responsible for the Vapor Compression Distillation Flight Experiment, which flew aboard the Shuttle on STS-107 in 2003 to demonstrate the use of a full-scale Urine Processor Assembly in weightlessness.

Other Areas of Responsibility

The Marshall Center also provides technical support for U.S.-supplied life support equipment onboard the Station, which, together with the Russian-supplied environment and life support equipment in the Zvezda service module, provide redundant systems for safety. This equipment, located in a payload rack in the U.S. Destiny laboratory module, Unity node, Airlock, and Multi-Purpose Logistics Module, includes:

- the Atmosphere Revitalization System and its components — the Trace Contaminant Control Subsystem, Major Constituent Analyzer, and Carbon Dioxide Removal Assembly — which removes carbon dioxide and trace contaminants from the cabin atmosphere and monitors the composition of the air;
- the Temperature and Humidity and Control subsystem, which helps maintain a habitable environment in the Station by removing heat and humidity, and circulating the cool dry air. Circulation of the atmosphere minimizes the temperature variations, ensures a well-mixed, breathable atmosphere and supports smoke detection;
- and the Fire Detection and Suppression subsystem, which provides fire detection sensors for the Station, fire extinguishers, portable breathing equipment and a system of alarms and automatic software actions to alert the crew and automatically respond to a fire.

Future

Ultimately, expendable life support equipment is not suitable for long duration, crewed missions away from low earth orbit due to the resupply requirements. It is expensive to continue launching fresh supplies of air, water and expendable life support equipment to the Station and returning used equipment to Earth. On deep space missions in the future, such resupply will not be possible due to the distances involved, and it will not be possible to take along all the water and air required due to the volume and mass of consumables required for a voyage of months or years. Regenerative life support hardware, which can be used repeatedly to generate and recycle the life sustaining elements required by human travelers, is essential for long duration trips into space.

National Aeronautics and Space Administration

George C. Marshall Space Flight Center
Huntsville, AL 35812
www.nasa.gov/marshall

www.nasa.gov

FS-2008-05-83-MSFC
8-368788

Appendix 2: NASA's Current Plans for Human Exploration Above Low Earth Orbit Include the Multi-Purpose Crewed Vehicle (Orion) and the Huge Space Launch System (SLS) Booster

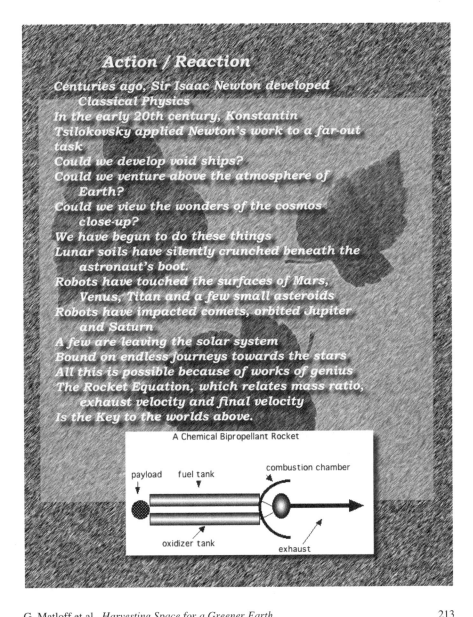

Action / Reaction

Centuries ago, Sir Isaac Newton developed
    Classical Physics
In the early 20th century, Konstantin
Tsilokovsky applied Newton's work to a far-out
task
Could we develop void ships?
Could we venture above the atmosphere of
    Earth?
Could we view the wonders of the cosmos
    close-up?
We have begun to do these things
Lunar soils have silently crunched beneath the
    astronaut's boot.
Robots have touched the surfaces of Mars,
    Venus, Titan and a few small asteroids
Robots have impacted comets, orbited Jupiter
    and Saturn
A few are leaving the solar system
Bound on endless journeys towards the stars
All this is possible because of works of genius
The Rocket Equation, which relates mass ratio,
    exhaust velocity and final velocity
Is the Key to the worlds above.

A Chemical Bipropellant Rocket

payload   fuel tank   combustion chamber

oxidizer tank   exhaust

G. Matloff et al., *Harvesting Space for a Greener Earth*,
DOI 10.1007/978-1-4614-9426-3, © Springer Science+Business Media New York 2014

$MR=Exp(delta V/Vex)$

Surface-Running Torpedo

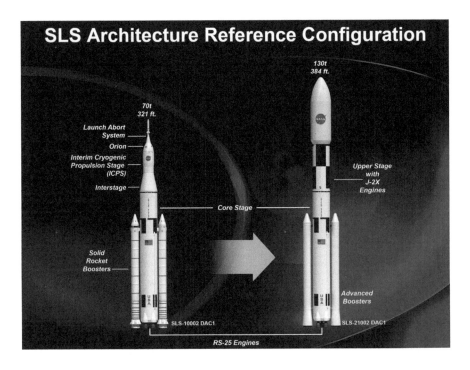

National Aeronautics and Space Administration

orion Quick Facts

Orion is America's next generation spacecraft that will take astronauts to exciting destinations never explored by humans. It will serve as the exploration vehicle that will carry the crew to distant planetary bodies, provide emergency abort capability, sustain the crew during space travel, and provide safe re-entry from deep space.

NASAfacts

52'11"

Orion Summary

Number of crew 4
Crewed mission duration .... 21-210 days
Total change in velocity .... 4920 ft/s
Gross liftoff weight .... 69,181 lbs
Effective mass to orbit .... 50,231 lbs

17'7.5"

Launch Abort System

Launch Abort System – Emergency Crew Escape System During Launch

Mass Properties

Dry mass/propellant .... 10,369 lbs
Gross liftoff weight .... 16,125 lbs

10'11"

16'5.5"

Crew Module

Crew Module – Crew and Cargo Transport

Pressurized volume (total) .... 690.6 ft³
Habitable volume (net) .... 316 ft³
Reaction control system (RCS) engine thrust .... 160 lbf/engine
Return payload .... 220 lbs

Mass Properties

Dry mass/propellant .... 21,350 lbs
Oxygen/nitrogen/water .... 77 lbs
Landing weight .... 19,463 lbs
Gross liftoff weight .... 21,650 lbs

18'4"

Service Module – Propulsion, Electrical Power, Fluids Storage

Mass Properties

Oxygen/nitrogen/water .... 694 lbs
Propellant weight .... 17,433 lbs
Gross liftoff weight .... 27,198 lbs

6'1"
13'10"

Service Module

The Orion Spacecraft

Launch Abort System
The Launch Abort System, positioned on a tower atop the crew module, can activate within milliseconds to pull the crew to safety and position the module for a safe landing.

Crew Module
The Crew Module (CM) is capable of transporting four crew members beyond low Earth orbit, providing a safe habitat from launch through landing and recovery.

Attitude Control Motor

Jettison Motor

Abort Motor

Fillet

Ogive Fairing

Docking Adapter

CM/SM Umbilical

Roll Control Thrusters

Orbital Maneuvering Engine

Service Module
The Service Module (SM) provides support to the CM from launch through CM separation prior to entry. It provides in-space propulsion capability for orbital transfer, attitude control, and high altitude ascent aborts. While mated with the CM, it also provides water, oxygen and nitrogen to support the CM living environment, generates and stores power while on orbit, and provides primary thermal control. The SM also has the capability to accommodate unpressurized cargo.

Solar Array

Spacecraft Adapter
The spacecraft adapter connects Orion to the Launch Vehicle.

National Aeronautics and Space Administration

Lyndon B. Johnson Space Center
Houston, Texas 77058

www.nasa.gov

FS-2011-12-058-JSC

Appendix 3: Mitigating Global Warming Using Ground-Based (Terrestrial) Geoengineering

Robert G Kennedy III, PE; Kenneth I. Roy, PE; Eric Hughes,
David E. Fields, Ph.D.

Acronyms

CCS	Carbon capture and sequestration
CDM/CDR/ CDA	Carbon dioxide management/removal/avoidance
CPI	Consumer Price Index
EHV	Extremely high voltage
EJ	Exajoule, large SI unit for energy $= 10^{18}$ J
ERO(E)I	Energy-return-on-(energy)-invested
GHG	Greenhouse gas
IPCC	Intergovernmental Panel on Climate Change
PV	Photovoltaic
Q	"Quad", large US customary unit for energy $= 10^{15}$ quadrillion British thermal units
SLOC	Sea line of communication
SRM	Solar radiation management
WWI/II	World Wars I and II

Introduction to Appendix 3

Next to escaping pathogen load and the burden of disease, environmental and resource crises have been the greatest strategic motivators for human exploration and migration into unpeopled realms. We cannot qualify this statement with "in human history", since most of these paleolithic peregrinations in search of epidemiological liberation took place tens of millennia ago, lost in the dustbin of prehistory. Even absent writing, the archaeological record demonstrates well

G. Matloff et al., *Harvesting Space for a Greener Earth*,
DOI 10.1007/978-1-4614-9426-3, © Springer Science+Business Media New York 2014

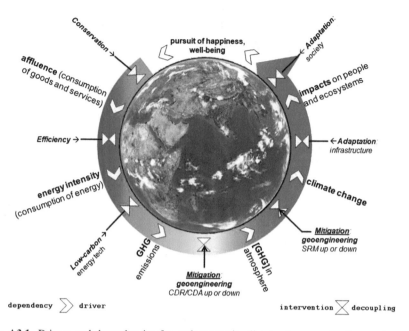

Fig. A3.1 Drivers and dependencies for anthropogenic climate change, with the spectrum of approaches and points of their application (emphasizing geoengineering) for intervention. Diagram by authors (2013), after Caldeira (2013)

enough that ancient societies that outgrew their *contado* (geographic extent of their resource base, typically fuel wood) crashed. So it is today—the climate crisis is upon us. We are still in the early exponential growth phases of it—the worst positive feedback loops (next section) have yet to commence. Wise guys have quipped, "we are *venusforming Terra*". It is the mission of this book to show that the climate challenge is in fact pertinent to the development of a spacefaring civilization—*and the other way round*! As the astronomy community realized after discovering the true cause of the dinosaurs' demise just 33 years ago, the ability to work in space is not a luxury, but a *necessity* for long-term survival. It is time for people to find a new *contado* by looking in a new direction—not to the far horizon, but up.

The Climate-Carbon-Water-Energy Nexus

Water means life, and we live on a water world. Ultimately, climate change (Fig. A3.1) is about water also: too much water where it is not supposed to be or when; not enough water where or when it is supposed to be; plus drying soils, changing habitats and ocean currents, and rising sea levels over the next century submerging the most historically important, densely occupied, and productive real

Fig. A3.2 A simple syllogism: one latent positive-feedback loop waiting to take off

CO₂ released ↱ wildfires / Temp. rise / eco stress / tree death

estate on Earth. Warm air holds much more water than cool air, and when it is released, weather events tend to be more violent and shift toward the extreme ends of the probability distribution. Agricultural practices and our modest pantry of staple crops (not to mention the wider ecosystem) are adapted to long term averages with predictable error bars. As the old saying goes, climate is the weather one expects, while weather is the precipitation one gets, therefore climate change means *getting the unexpected*. The greatest changes will occur at the poles, so nations of the far north which blithely boast that they will benefit from climate change when the temperate belt migrates north ought to reconsider this Panglossian proposition. For one thing, the extreme seasonal variation in day length up north due to Earth's tilt will not change just because it gets warmer. For another, soils of the north tend to be quite different in character (*podzols*) and fertility than the rich black earth (*chernozem*) and loess of the world's breadbaskets. As Jared Diamond pointed out in _Guns, Germs, and Steel_ (1997), it is far more difficult to shift a given crop's cultivation zone north or south across latitudes than it is east or west across longitudes. Other nations in northern Europe may actually get colder if the thermohaline current gets interrupted. Human activity to date has already left traces that will be legible in the geologic record eons hence, thus the term "Anthropocene" was coined. In effect, we are already conducting a planetary- experiment scale (albeit uncontrolled) in geoengineering. "Mitigation" is what we must do to delay or prevent climate change. "Adaptation" is what we have to do after it has happened. Geoengineering, then, belongs among mitigation strategies. Since the IPCC's _Fourth Report_ in 2007, it has become clear that a certain amount of global warming is already "dialed in" now, no matter what we do next. Therefore, coping with climate change means *managing the unavoidable*, while *avoiding the unmanageable*.

There are three variables in global warming:

1. The first is the concentration in the atmosphere of greenhouse gases (GHG), e.g., water vapor (H_2O), carbon dioxide (CO_2), methane (CH_4), plus other minor ones. Since the most powerful GHG by sheer quantity is water vapor, which is produced by solar heat evaporating liquid water, one can appreciate that gaseous H_2O is the basis for a quick-response positive-feedback temperature loop. CO_2 is the basis of a delayed-response positive feedback temperature loop (Fig. A3.2), via release from burning forests killed by climate stress—boreal are the most vulnerable—or by direct release from warming permafrost. The carbon load in a mature forest is ~1 g/cm^2, or 40 tons per acre—times several billion boreal acres makes an immense amount of standing carbon waiting for a match. CH_4 can also be the basis of such a loop when warming permafrost directly releases methane. The sum of GHG frozen in the tundra is several times what is in the entire atmosphere already. The trouble with a positive-feedback loop is that, once

initiated, it tends to run to completion, so it is something strongly to avoid if you want to engineer a controllable system.

2. The second is the quantity of light of summed across all wavelengths continuously coming in from the Sun, also known as "the radiative forcing function".

3. The third is how much incoming energy is reflected immediately, or *albedo*. Dark surfaces (e.g., water) absorb energy; light ones (e.g., ice) reflect it. Sunlight directly warms our planet because our atmosphere is transparent to the predominant wavelengths of solar radiation: visible light.

But GHGs in our air partly block the IR re-radiating from the warmed surface, trapping some heat. This greenhouse effect keeps the overall planet warm (even at night) making it possible for life to exist. Without a blanket of GHGs, plus lots of that truly amazing substance, liquid water, to moderate temperature extremes, Earth would be a cold, inhospitable place.

Humanity's efforts are currently focused (unsuccessfully) on reducing the GHGs (principally CO_2) being emitted to the atmosphere, mainly via (ineffective) top-down attempts to limit the consumption of carbonaceous fossil fuels. Conservation, recycling, more efficiency, altered lifestyles, and new low-carbon sources of energy are all needed to reduce our GHG emissions. But so far, all that these tactics have achieved is reducing the *rate* of emission *growth*, and not even that in most places. Absolute consumption of fuel, and concomitant emissions and concentration [CO_2] in the atmosphere), continues to rise dramatically. To avoid the worst effects of climate change, humanity needs to do more than merely reduce growth; we must bring down the absolute amount of these gases in the atmosphere to levels below those of the last century. In the advanced tertiary and quaternary economies of the developed world, particularly the USA, economic growth is noticeably decoupling from energy growth, albeit still on a very high base of consumption. Not so in the developing world—with more people (numbers, N) entering the global middle class (affluence, A) with commensurate use of more energy per capita yet nearly flat energy intensity (I) in the developing world, it does not seem we will be able to reduce the footprint of the gross energy product, $E = N \times A \times I$, and stop global warming. With $\sim 10^{10}$ mouths to feed on this world by mid-century (recently revised upward +1 billion by UN demographers), we certainly will not be able to conserve our way to pre-twentieth-century GHG concentrations—not without falling back to pre-twentieth-century living standards at any rate. Who would willingly consent to that? Who or what authority would enforce it? Abstinence is a personal choice, not a viable policy of first resort.

Fully oxidizing a carbon atom turns 12 Da of fuel into 44 Da of product, a 3.67:1 ratio. This means that the ~10 gigatonnes (Gt) of stored sunlight (fossil fuel) that we dig out of the Earth every year gets quadrupled to ~40 gigatonnes of pollution (and then two-thirds of that energy is wasted immediately). While this 40 Gt/year is only ~10 % of the natural 400 Gt/year total flow of into CO_2 and out of the world's forests and soils, those latter flows are balanced. Ours are not, and this is the key difference. Humanity's GHG pulse is always positive, and growing exponentially, driven by the factors illustrated in Fig. A3.1 above.

The sheer scales of the coupled energy and climate problems dwarf those of the Manhattan Project and Apollo programs, oft cited by innumerate politicos. These historic top-down exhibitions of the State's power are not even a rounding error compared to petroleum—just by itself is the largest industry in the world, $3–5 trillion per year. The oil business is ~5¢ of every dollar on Earth, every year, depending on crude oil prices. (Oil subsidies in the developing world are an even larger slice of the national pie.) Since this is a space book, we offer the following statistic for the reader's consideration: the gasoline habit of just one typical large Western city in one day, ~5 <u>kilotonnes</u>, outweighs the sum of all payload mass (not the launchers or fuel) shot into space since October 4, 1957, and still up there today.

Despite these seemingly imposing numbers, know that one hour of sunshine falling on the lighted face of Earth is about the same as the primary energy budget of the entire human race today, ~500 quads per year. (A "quad" stands for "**quad**rillion British thermal units", i.e., 1,000,000,000,000,000 Btu. This is a very large unit by human standards, equal to the yield of 250 megaton-range hydrogen bombs.) 500 quads certainly looks like a lot, but it pales in comparison to what our star sheds on our little world, an 8766-to-1 ratio. Furthermore cost of photovoltaic (PV) technology has reliably fallen by an average of 12 % per year for the last 20 years, while at the same time PV is fastest growing form of all generating capacity, with installations up +33 % to +50 % per year.

Three major revolutions are going on in the energy domain right now. The first is the extraordinary phenomenon of cheap ubiquitous natural gas, liberated from deep shale by the unforeseen combination of horizontal drilling, hydrofracking, and slippery frac fluids, and rapidly transitioning to a truly global fuel thanks to virtual transoceanic pipelines, i.e., LNG tankers. This is having geopolitical consequences. As DeTocqueville might have said, it has "grown up unnoticed...whilst the attention of mankind was directed elsewhere". The second is the proliferation of PV, thanks to the unintended consequence of European feed-in-tariffs stimulating the Chinese PV industry to grow like gangbusters. The third is the coming end of oil exports from traditional producing nations by the mid-2030s, as their internal populations' consumption absorbs more of their own oil resource. Two other remarkable economic phenomena to note: for the past 40 years, retail electricity in America has appreciated at only ~2 % per year, which is less than the annual consumer price index (CPI) of ~3 %. In other words, electricity actually gets cheaper over time in constant dollars than almost everything else! Over the same period, petroleum-derived motor fuels such as #2 diesel have appreciated at more than twice the CPI and triple the electricity index! Which long-term pricing trend would you rather ride?

Figure A3.3 below shows that demand for electricity, the highest, most useful form of energy, is growing faster than any other. Electricity comprises ~1-part-in-8 of primary energy consumption today, but the world will be much more wired by midcentury, ~1-part-in-5, thanks to differential compound growth. Nameplate electric generating capacity is on a path to triple to ~10 TW_e by 2050. The world's electricity grid is worth ~$20 trillion today (average overnight capital cost of $4/W, ~3.5 terawatts total nameplate capacity, 40-year useful life). Global generation,

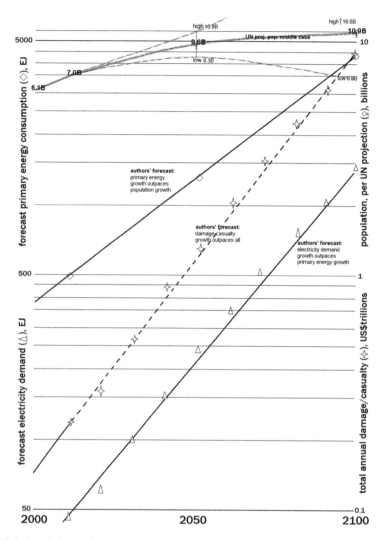

Fig. A3.3 Population projections by the United Nations' demographers (top curve), low-middle-high cases, plotted vs. authors' forecasts for annual climate-related damage, primary energy consumption, and electricity demand, for 2000–2100. 1 exajoule (SI symbol: EJ, 10^{18} J) ≈ 1 quadrillion Btu (symbol: Q, 10^{15} Btu). Note logarithmic scale. Population is increasing by ~1 %, primary energy by ~2 %, and electricity by ~3 % *per annum*. Primary energy demand today is ~500 quadrillion Btu/year (~17 TW_t), while steady-state (24/7) demand electricity is ~2 TW_e

transmission, and distribution of electricity is already a trillion-dollar annual enterprise (about the same scope as the worldwide automobile industry). Thus, it is self-evident that tens of trillions of dollars of capital expense (CapEx) is *already* programmed to maintain, replace, and expand the grid over the next four decades.

This is a river of cash flow to organically harness without invoking top-down crash programs by *fiat*.

Also in Fig. A3.3, we project the burden of damage related to climate change to rise from 0.1 % of global product now to 1 % by 2100, while the global economic product itself grows at a compound annual rate of about 3 % (not plotted, but highly correlated with primary energy in the developing world). The world's economy grew at a real annual rate of 2 % in the period 1870–1914, a multipolar era that was much like the present one, and 4 % during the period 1957–2007. So 3 % is a reasonable century-long average. This forecast economic damage is driven by three heterodyning trends:

1. More people are dwelling within, or moving into, more marginal lands, which are more exposed to climate-change risk;
2. A more tightly-coupled global economy transmits damage signals better throughout society; and
3. In an economic analogue of Metcalfe's Law of networks, the more people there are, the more any given material thing exposed to the risk is worth.

Casualty trends since 1950 suggest we are already on this (ill-)starred path. The sum of these expected casualties over the coming century is on the close order of US$200 trillion in nominal dollars; the present value is $20–50 trillion, discounted to 2010 dollars. Just avoiding this damage compared to business-as-usual is likely worth more than the cost of whatever countermeasures we devise, but only to the degree that those measures take effect in time and scale in proportion to the problem. This is the shaky value proposition.

Traditional risk analysis (and literally shortsighted economic discounting of large, but far off events) fails in such situations because "Black Swans" are more common than people living comfortable lives in the developed world realize. Traditional economic analysis falls literally short (-sighted) because typical discount rates diminish the present value of a benefit (or a loss) beyond a generation in the future to virtually nothing. Survival considerations trump the ordinary calculus of decision-making, but no conventional equation can "prove" it. (Hence the aphorism, "*National security has no price*.") If the causes of climate change cannot be eliminated, can we do anything to mitigate it?

A Taxonomy of Terrestrial Geoengineering

Planetary engineering, or *geoengineering* (as coined by Marchetti in 1976 in the context of terrestrial carbon sequestration), is now understood to be the *intentional* application of technology, either ground- or space-based, for the purpose of influencing the properties of a planet on a global scale. No study of coping with climate change is complete without considering it.

In a comprehensive taxonomy, the gamut of geoengineering techniques can be cleaved along one axis by _locus_, i.e., ground-based and space-based, as in the top and bottom of Fig. A3.4 below. Geoengineering techniques can also be sorted on another, orthogonal axis, according to _approach_, in which the same elements are separated into two broad categories: _solar radiation_ management (SRM, left side) and _carbon dioxide_ management (CDM, right side), which is comprised of removal (CDR) as well as avoidance (CDA). (CDR in turn contains both biologically-based methods as well as chemical/industrial ones, which includes the subset of carbon capture and sequestration, CCS.) Within each of the resulting four quadrants, techniques can be further distinguished by another echelon of locus in order of increasing altitude, viz., sea-land-air on Earth, or LEO-GEO-SEL1 (Sun-Earth L1 point) in space. In this chapter, we will start on the ground level by discussing the lower half of the chart, while the upper half will be the subject of the next.

Some of the techniques are probably effective enough (denoted by **boldface**) to produce measurable effects, quickly (denoted by _italics_) and cheap enough (dimension not charted, but mere low \$billions/year) to be within the reach of individual nation states or even wealthy individuals. This single realization, which only dawned on the Establishment perhaps 3 years ago, is what has brought the subject of geoengineering from out of the ghetto of science fiction into the _fora_ of serious national policymaking in the USA. This debate is not isolated in the halls of Congress, either—it is also taking place behind the walls of the Kremlin. Having executed an about face and summarily declared as a matter of state doctrine that climate change does indeed exist (a remarkable development for a nation so wedded to the fossil fuel economy and controlled by so few individuals), Russian science is now seriously weighing the efficacy of terrestrial versus space-based approaches. Geoengineering is becoming a salient issue in other places as well. Most of the ground-based methods either require continual expenditure of effort for the benefit, or do not scale well, or are fraught with major downsides such as acid rain and other ecological side-effects, unintended consequences, or irreversibility, as well as moral hazards. While they would neither be precluded nor contravene existing international law regarding environmental modification ("ENMOD"), since the 1976 Treaty is explicitly predicated on "hostile use", these risky measures do seem suitable for use "only in uttermost need".

Taxonomy in words. Summaries of each of the ground-based geoengineering techniques follow, listed in the same vertical order as they appear in Fig. A3.4. The more notable items in the list are denoted by **boldface** (Table A3.1).

– Spectrum of the Terrestrial SRM quadrant (bottom of Fig. 14.1):

- _**Modifying Earth's albedo by injecting manmade sulfate aerosols into the**_
  **stratosphere.** (Budyko, 1974; Izrael, 2005, etc.) This approach artificially
  mimics the known after effects of major volcanic eruptions, in which sunlight
  is reflected by the suspended particles. Sulfate aerosols can be dispensed from
  special-purpose high-flying aircraft or by very large power plants with unusually tall smokestacks burning high-sulfur ("dirty") coal. (Mt. Pinatubo in
  1991 produced an immediate cooling effect over the globe, averaging

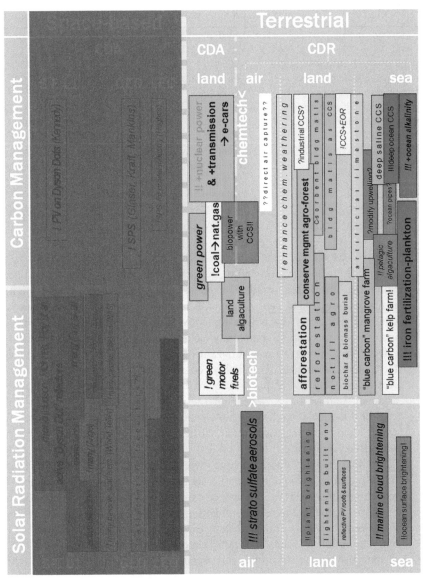

To ensure legibility despite no icons, simple typographical tricks are utilized in this chart as cues to convey six dimensions of data from this mere two-dimensional sheet of paper. **Cost** is not charted. Key:

Effect: tiny 8 pt.=insignificant; small 10 pt.=minor; plain 12 pt.=significant; large 14 pt. **bold=major**; see Table 14-1 next page for numerical values of Effect

Speed/Response: e x p a n d e d   t e x t=very slow or slow (millennia to centuries); plain=multiple decades; *condensed italic=quick (years)*

Risk: ?=unknown; green=none/safe; !yellow=low; !!orange=moderate; !!!red=high/dangerous

Fig. A3.4 A taxonomy of *terrestrial* geoengineering approaches comparing useful effect/scale, speed, and risk. See summaries next

Table A3.1 Terms of assessment and what these evaluations mean numerically or qualitatively

Cost (not shown Fig. 14.4) assessment term	"Negligible"	"Low"	"Modest"	"High"
What it means (US$billions/year)	<1	<10	<100	1000+
Assessment of scale/effectiveness	"Insignificant"	"Minor"	"Significant"	"Major"
What it means (gigatonnes/year)	<0.3	~1	~3	10+
Approx. forcing function equiv. (W/m^2)	<<0.1	0.1	0.3	1+
Speed/response time assessment term	"Very slow"	"Slow"	"Half-fast"	"Quick"
What it means	Millennia	Centuries	Decades	Years
Risk assessment term	"None"	"Low"	"Moderate"	"High"
What does it mean?	Not zero but safe	Think first	Sporty game	Dangerous
Qualifiers	"Unknown"	"Possibly"	"Probably"	*No word*
What they Mean	*Unable to assess*	*<30 %*	*>70 %*	*=Certain*

−0.5 °C, which would have been −3 °C if the dust didn't fall out so fast.) Downsides include irreversibility, acid rain, and increased oceanic acidification. Cost: **low**. Effect: **major**. Response: **quick** (incremental). Risk: **high**.

- Brightening eco landscape to increase reflectance by cultivating species of wild plants or crops selected for higher overall albedo than others. Downsides include site specificity, complexity/lack of general applicability or uniform effect, irreversibility, invasive species issues, altered wildlife habitat and feed, increased fuel load (enthalpy) contributing to increased wildfire risk. Any technique based on plants has two fundamental disadvantages: there 3× as much non-land (i.e. water) on the globe as dry land, of which only one-third is arable; to thrive, plants generally require a stable environment, which contradicts the assumption. Cost: **modest**. Effect: **minor**. Response: **slow**. Risk: **moderate**.

- Lightening the built environment to increase solar reflectance by white roofs, roads, etc. which would also ameliorate local "heat island" effect, somewhat reducing energy for HVAC. May be cost-effective if integrated with program of normal scheduled replacement. EROI: modest. Downside if applied faster than normal includes misallocated resources, suboptimal investment. The built environment (roads and roofs) covers perhaps 100 million acres (5 %) of the USA's total 2 billion acres; scaling from GDP suggests worldwide coverage is 300 million acres at most, only ~0.2 % of global surface area. "Signal would be 3 dB below noise." Cost: **low to high**. Effect: **insignificant**. Response: **very slow**. Risk: **none**.

- PV roofs and built surfaces to increase all-wave reflectance by locating distributed power and heat generation on roofs, roads, etc. which offsets avoids GHG emission and rejected heat from traditional thermal power cycles, plus reduces "heat island" and local HVAC loads. Subset of above: only ~0.2 % of global surface area if built out but synergistic benefits. Cost-effective if integrated with normal scheduled replacement. EROI: high. Cost:

low to modest. Effect: **minor to possibly significant**. Response: **half-fast**. Risk: **none**.

- *Modifying Earth's albedo by artificially generating bright marine clouds* (Gadian, Salter, 2011). Large special-purpose vessels in midocean atomize salt spray and inject it into upper air. More and brighter clouds of greater persistence reflect more incident sunlight than ordinary clouds over land. Downsides unknown but could include disruption of rainfall patterns over land in time and space domains. Probably reversible. Cost: **low**. Effect: **significant to major**. Response: **quick** (incremental). Risk: **moderate**.

- Lightening the ocean's surface to increase solar reflectance by floating light-colored particles or mats. (Seitz) Downsides include interference or hazard with maritime navigation, interference with marine aquatic and avian wild-life, indistinguishable from solid waste pollution. Cost: **modest**. Effect: **minor to significant**. Response: **slow**. Risk: **moderate**.

- Bag o'tricks in the Terrestrial CDM quadrant, CDA section (upper middle of Fig. A3.4):

  - *Green power (solar, geothermal, wind)*. 95+ % decarbonization per unit of electricity produced. Distributed generation leads to more secure governance. Solar-electric peaking power and geothermal baseload electricity are complementary. Cost-effective if integrated with program of normal scheduled replacement. Build-out would be self-funding, synergistic with extension of transmission network at continental scale would synergize build-out of wind and solar power far from load centers. Stepping stone toward Global Grid. PV is fastest growing form of generation, installations up +33 % to +50 % annually. Second of three major revolutions in energy domain. Market space up to 1 TW in near future, up to ~3 TW by midcentury. Potential annual GHG reduction by midcentury ~10+ Gt/year. Consumer financial savings would be indirect due to isolation from external price volatility. High EROI. Downsides include increased land use, complexity, integrating intermittent power into grid. Cost: **modest to high**. Effect: **major**. Response: **half-fast**. Risk: **none**.

  - Shift electricity generation from coal to natural gas. CH_4 has half the carbon of coal, per unit weight; and gas-fired combined-cycle plants operate at almost double the thermodynamic efficiency of conventional coal-fired power plants; chained improvement, $4\times$, or -75 % decarbonization per unit of electricity produced. Cost-effective if integrated with program of normal scheduled replacement. Proliferation of cheap natural gas from fracked deep shale is the first of three major revolutions in energy domain. Consumer financial savings would be indirect but substantial due to temporary autarky/liberation from imported petroleum as disruption of cartels, rentierist-state-owned oil companies. Potential annual GHG reduction by midcentury ~6 Gt/year. Downsides include medium-term "bridge" fuel, not permanent solution; pipeline accidents, fugitive CH_4 emissions (powerful

GHG), blocking effect on nuclear and green power, societal slacking off. Cost: **modest**. Effect: **significant to major**. Response: **half-fast**. Risk: **low**.

- **Electrifying half the automotive fleet with low-carbon nuclear power and extremely high voltage transmission network at continental scale**. This would eliminate ~5 Gt/year of GHG emission now, 12 % of present world total. Cost-effective if integrated with program of normal scheduled replacement. Consumer financial savings would be substantial, several $trillion per year. Extra transmission capacity is pennies on the dollar compared to extra generating capacity. Build-out would be self-funding, extending network at continental scale would synergize build-out of green power far from load centers. Good EROI. Stepping stone toward Global Grid (aka Dymaxion Grid, see below). Downsides include nuclear reactor accidents due to operation by non-learning organizations in opaque societies. Cost: **modest to high**. Effect: **significant to major**. Response: **half-fast**. Risk: **low to moderate**.

- **Green motor fuels**. Assuming carbon-neutral production based on renewable feedstocks, this would neutralize up to another ~5 Gt/year of fossil GHG emission, 12 % of present world total. Consumer financial savings would be substantial, a $trillion or two per year. EROI highly dependent on feedstock choices, e.g., cellulosic vs. algal, corn vs. sugar. EROI for corn ethanol is negative to barely fractionally positive. Downsides include land use, water use, potential competition with food. Cost: **modest to high**. Effect: **significant to major**. Response: **half-fast**. Risk: **low**.

- Biopower with CCS. Renewable biomass-fired power plants with oxyfiring, exhaust captured and sequestered. Hybrid CDR/CDA method. CCS is unproven at acceptable parasitic load at industrial scale; storage methods of uncertain longevity. CCS transport infrastructure would have to be at least as great as entire gas+oil+coal networks now, worth ~$10 trillion worldwide, just to handle necessary mass flow. Downsides include pressure on forests especially overseas, marginal EROI, leaky geologic storage. Cost: **modest to high**. Effect: **minor**. Response: **slow**. Risk: **moderate**.

– Portfolio of approaches in the Terrestrial CDM quadrant, CDR section (top of Fig. A3.4, right side):

- Direct-air CCS is unproven technology. Operational and capital ROI unknown. Potential net annual GHG absorption by midcentury: unknown, probably minor. Location-neutral. Possibly high cost for CO_2 transport infrastructure. EROI: unknown. Cost: **possibly high**. Effect: **probably insignificant**. Response: **probably slow**. Risk: **unknown**.

- Enhanced chemical weathering. Inject large volumes of engineered nanoscale dust particles into the atmosphere to absorb carbon, form carbonates, mimicking natural dust storms and weathering. Large surface area per unit mass for CO_2 to adsorb onto, compared to rocks and building/road surfaces. Stability: probably good, chemical reaction in air self-limiting. Technology: probably feasible. Scalability: uncertain. Downsides: indistinguishable from pollution, health hazards of inhaled dust on people and wildlife, increased maintenance on buildings, accelerated wear on machines. Cost: **modest to high**. Effect: **minor to significant**. Response: (extremely) **slow**. Risk: **low to moderate**.

- No-till agriculture preserves the underground soil-food-web by not disrupting it, as well as keeping the soil from drying out and blowing away. The subsurface contains perhaps 30 % of the biomass standing above it. This is both a removal and an avoidance strategy. Actively farmed land in the USA covers perhaps 200 million acres (10 %) of total; pasturage is another 10 %, with standing carbon loads anywhere between 1 and 10 tons/acre. Scaling from USA suggests worldwide coverage is over 2 billion acres, over 1 % of global surface area; contains ~6 Gt standing carbon, meaning 2 Gt subsurface biota. No-till agriculture also consumes much less energy, mainly motor fuel, so this option may produce net savings. *Marginal* cost: **negligible to low**. Effect: **minor to significant**. Response: **half-fast**. Risk: **none**.

- Afforestation by planting trees on marginal land that was not forested before. Candidate land in the USA covers perhaps 600 million acres (30 %) of total. Scaling from USA suggests worldwide coverage is almost 3 billion acres, over ~2 % of global surface area; could absorb 10 Gt/year initially during early successional stage before stabilizing at 3 Gt/year absorption in maturity; eventually contain 120+ Gt standing carbon, plus 30 % subsurface biota. Considerable upsides: net new wildlife habitat, renewable materials, silviculture, runoff/water filtration/cleansing. Strong net carbon reduction. Downsides include typical risks if monoculture, site specificity, invasive species issues, increased fuel load (enthalpy) contributing to increased wildfire risk. Furthermore, any technique based on plants has two fundamental disadvantages: there is three times as much non-land (i.e. water) on the globe as dry land, of which one-third is arable; to thrive, forests generally require a long-term stable environment, which contradicts the assumption. Cost: **high**. Effect: **major**. Response: **initially quick, then half-fast**. Risk: (very) **low**.

- Reforestation by re-planting trees on cleared land that was once forested. Candidate land in the USA covers perhaps 200 million acres (10 %) of total. Scaling from USA suggests worldwide coverage is over a billion acres, ~1 % of global surface area; could absorb 2 Gt/year initially during early successional stage before stabilizing at <1 Gt/year absorption in maturity; eventually contain ~30 Gt standing carbon, plus 30 % subsurface biota. Considerable upsides: wildlife habitat, renewable materials, silviculture, runoff/water filtration/cleansing. Better if forest had not been disturbed in the first place. Downsides include typical risks if monoculture, site specificity, invasive species issues, increased fuel load (enthalpy) contributing to increased wildfire risk. Furthermore, any technique based on plants has two fundamental disadvantages: there is three times as much non-land (i.e. water) on the globe as dry land, of which one-third is arable; to thrive, forests generally require a long-term environment, which contradicts the assumption.

Cost: **modest**. Effect: **significant**. Response: **initially quick, then half-fast**.
Risk: **none**.

- Forest conservation & management. 1 g/cm^2, compare to 1 kg/cm^2 of air. Not disturbing a mature forest in the first place which is already adapted to its location keeps 10–40 tons of carbon per acre in place bound up as biomass, plus an additional 3–10 tons per acre subsurface biota. *Marginal* cost: **negligible to low**. Effect: **major**. Response: *not applicable*. Risk: **none**.

- Select building materials and surfaces in the built environment to absorb carbon. Stability: uncertain. Technology: uncertain. Scope of solution unmatched to problem, must be part of portfolio if done at all. Possibly cost-effective if integrated with program of normal scheduled replacement. Downside if applied faster includes misallocated resources, suboptimal investment. The built environment (roads and roofs) covers perhaps 100 million acres (5 %) of the USA's total 2 billion acres; scaling from GDP suggests worldwide coverage is 300 million acres at most, only ~0.2 % of global surface area. "Signal would be 3 dB below noise." Cost: **unknown**. Effect: **insignificant**. Response: (very) **slow**. Risk: **none**.

- Select building materials that contain sequestered carbon, e.g. non-Portland cement. Stability: secure. Scope of solution unmatched to problem, must be part of portfolio if done at all. Possibly cost-effective if integrated with program of normal scheduled replacement. Downside if applied faster includes misallocated resources, suboptimal investment. The built environment (roads and roofs) covers perhaps 100 million acres (5 %) of the USA's total 2 billion acres; scaling from GDP suggests worldwide coverage is 300 million acres at most, only ~0.2 % of global surface area. "Signal would be 3 dB below noise." Cost: **unknown**. Effect: **insignificant**. Response: (very) **slow**. Risk: **none**.

- Biochar and biomass burial in underground caverns or excavations. Potential annual GHG absorption by midcentury: unknown, probably <Gt/year. Geologic stability: secure. Feasibility highly location-specific. EROI: probably marginal. Downsides include moderate to high cost for handling and transport. Cost: **modest**. Effect: **minor**. Response: **slow**. Risk: **none**.

- CCS in deep saline aquifers. Aqueous CO_2 sequestered in deep underground saline aquifers. Potential annual GHG absorption by midcentury: unknown, possibly Gt/year. Geologic stability: unknown, but probably secure. Feasibility highly location-specific. EROI: n/a. Downsides include moderate to high cost for transport infrastructure. Cost: **low**. Effect: **possibly significant**. Response: **half-fast**. Risk: **none**.

- CCS+EOR. Supercritical CO_2 utilized for enhanced oil recovery, then sequestered in oil fields or saline aquifers. Possibly carbon-neutral now; potential net annual GHG absorption by midcentury: unknown, probably minor. Geologic stability: highly location-specific. High cost for pipeline transport infrastructure, but probably self-funding, even profitable. EROI: unknown. Leak issue moot since CO_2 source is fugitive in the first place. Downsides include moderate to high cost for transport infrastructure and ocean acidification, esp. local. Cost: **modest**. Effect: **probably minor**. Response: **quick**. Risk: **low**.

- Accelerated limestone deposition. Engineer matrices and flows to form solid carbonates in either salt of freshwater bodies, utilizing either biological or synthetic chemical processes that mimics formation of natural shell, stalactite/stalagmites or hard water scale. Stability: probably good, chemical process is self-limiting. Technology: possibly feasible. Scalability: uncertain. Cost: **unknown**. Effect: **probably minor**. Response: (extremely) **slow**. Risk: **negligible to low**.

- "Blue carbon" mangrove farming and wetland creation/restoration. Cultivate mangrove/wetlands to absorb CO_2. Considerable upsides: complex wildlife habitat, storm surge barrier, runoff/water filtration/cleansing. Highly productive tropical ecosystem absorbs up to $10\times$ more carbon per acre per year than temperate forests and up to $100\times$ more carbon than boreal forests. Rapid maturity at similar but somewhat higher saturation of standing carbon. Potential annual GHG absorption by midcentury: unknown, but @ ~1+ ton/acre/year, possibly Gt/year. EROI: n/a. Permanence of sequestration unlikely. Downsides include typical monoculture risks, reservoir for disease, pests, or invasives. Cost: **negligible to low** (if part of long-term integrated management plan). Effect: **possibly significant to major**. Response: **probably slow**. Risk: **low**.

- "Blue carbon" pelagic kelp farming. Cultivate kelp to absorb CO_2 to be carried to depth in some form. Potential annual GHG absorption by midcentury: unknown, possibly Gt/year. EROI: n/a. Permanence of sequestration unlikely. Downsides include typical monoculture risks, amplifier for aquatic pathogens or invasives, disruption of ocean food web. Cost: **unknown**. Effect: **possibly minor to significant**. Response: **probably slow**. Risk: **low**.

- "Ocean pipes" (Lovelock, 2007). Artificial upwelling to bring cold nutrient-rich water to surface to stimulate plankton growth, absorption of CO_2. Not OTEC. Technology readiness uncertain. Potential annual GHG displacement by midcentury: unknown, possibly Gt/year. EROI: unknown. Downsides include mixing species among layers, unintended blooms followed by local eutrophication, aquatic life kill, intentional disruption of thermoclines, disruption of ocean food web. Cost: **unknown**. Effect: **possibly modest**. Response: **possibly quick**. Risk: **moderate to high**.

- Modify oceanic upwellings. Inject CO_2 in some form to be carried to depth. Potential annual GHG absorption by midcentury: unknown, possibly Gt/year. EROI: n/a. Technology readiness uncertain. Might be irreversible. Downsides include uncontained aquatic microbial invasive species, unintended blooms followed by local eutrophication, disease vectors, disruption of ocean food web. Cost: **unknown**. Effect: **possibly major**. Response: **possibly quick**. Risk: **moderate to high**.

- Pelagic algaculture. Algal feedstock for biofuels farmed in open ocean. Potential annual GHG absorption by midcentury: unknown, possibly Gt/year;

net emission if carbon-neutral: zero. Consumer savings, EROI: unknown. Might be irreversible. Downsides include uncontained aquatic microbial invasive species, disease vectors, disruption of ocean food web. Cost: **modest to high**. Effect: **possibly significant to major**. Response: **possibly quick**. Risk: **moderate to high**.

- Benthic CCS. Supercritical CO_2 sequestered in deep ocean layers or sediments. Potential annual GHG absorption by midcentury: unknown, possibly Gt/year. Geologic stability: unknown. EROI: n/a. Might be irreversible. Downsides include high cost for transport infrastructure and ocean acidification, esp. local; others unknown but could include alteration of buffering capacity, disruption of ocean food web, catastrophic release of gas. Cost: **high**. Effect: **possibly significant to major**. Response: **slow**. Risk: **high**.
- *Stimulating oceanic phytoplankton blooms with iron fertilizers*. (Bathmann, 2011) Fe filings dispensed from large special-purpose vessels in midocean. Potential annual GHG absorption by midcentury: unknown, possibly Gt/year. EROI: n/a. Might be irreversible. Downsides unknown but could include alteration of buffering capacity, disruption of ocean food web. Cost: **low**. Effect: **possibly major**. Response: **possibly quick**. Risk: **high**.
- *Reversing oceanic acidification by adding alkalinity a la ferro-fertilization above*. (Bathmann, 2011) Basic materials dispensed from large special-purpose vessels in midocean. Not a CO_2 absorption technique *per se*; potential annual GHG absorption by midcentury: n/a. EROI: n/a. Might be irreversible. Downsides unknown but could include alteration of buffering capacity, disruption of ocean food web, shell formation from diatoms to megafauna. Cost: **modest**. Effect: **possibly major**. Response: **possibly quick**. Risk: **high**.

Ruminations About Radiation Forcing Functions

SRM and a historical coincidence. Decreasing temperature by artificially increasing albedo *if done at sufficient scale* would almost immediately reduce the atmosphere's water vapor content, which should reverse the increasing frequency and severity of storms, limiting the damage from climate change. How much sunlight must be stopped to deal with climate change? The often-cited figure of $-2\ \%$ reduction in total insolation to achieve a designated temperature reduction was derived from a simple proportionality using the Stefan-Boltzmann law (σT^4) applied to average radiation over a spherical Earth. However, Earth is not a perfect blackbody, and her radiation budget across all wavelengths is not yet fully understood. In 2001, we conjectured that the adjustment to the radiative forcing function might be an order of magnitude more modest than 2 % based on the known effects of a widely observed natural astronomical event in human history, and a numerical coincidence with history, and suggested that a 0.25 % reduction of the insolation thru Earth's cross-sectional disk could still have a substantial beneficial effect.

To begin to bound the problem, recall that the sunspot cycle shut down for unknown reasons between the mid-sixteenth and seventeenth centuries (about the time of the discovery of their true nature, by Galileo among others). *Astronomers estimate that a quarter-percent (0.25 %) reduction in the Sol's luminosity accompanied this event*, and refer to the period as the *Maunder minimum*. Historians call it *"the Little Ice Age"*. Chronicles tell us that the Thames River in England froze for the first time in recorded history; sea ice cut off Iceland from Europe; crops failed. Parish records show European population growth stalled. Tycho Brahe (Kepler's mentor) recorded winter temperatures 2.7°Fahrenheit (1.5 °C) below average during the last two decades of the sixteenth century. Coincidentally, this is the same magnitude, but opposite sign, as the global warming forecast in the IPCC's *Third Report*. Having presented itself, we noted that a reduction of this close order in the radiative forcing function should offset that forecast warming. (*Nota bene*: while this approach could adjust the average global temperature, it would not address other environmental consequences of the continued burning of fossil fuels, such as acidification of the World Ocean. That must still be dealt with. Even though a technique conceptually similar to ferro-fertilization of phytoplankton has been proposed by the same geoengineers at the Wegener Institute to reverse acidification and increase the ocean's alkalinity by scattering basic particles, any "techno-fix" that does not change the original behavior which led to the problem by definition contains moral hazard.)

Either way, applying 2 % to the average insolation over the globe, or 0.25 % to the solar constant at Earth orbit, yields a similar figure: ~ -3.5 W/m^2. (We note that the current value is 1,361 W/m^2, but it really is not constant, as it consists of multiple superposed solar cycles.)

CDM. Removing sufficient CO_2 from the air, or avoiding its emission in the first place, eventually would also reduce the atmosphere's water vapor content, likewise limiting damage, though with some time delay. How much? Something like a teratonne of CO_2 must be taken out to restore us to a pre-Industrial Revolution concentration of GHG. But energy consumption is growing exponentially, on a path to triple by 2050—in just one short human generation, removing that same teratonne will only return us to the concentration of WWII! This is the counterintuitive bite of exponential growth phenomena, which human mental toolkits do not seem to be made to handle appropriately.

Figure A3.5 above depicts various natural and manmade radiative forcing functions from the IPCC *Fourth Report*. Most terrestrial approaches (in effect, *engineered* forcing functions of ~0.01–0.1 W/m^2) do not scale to the problem, which calls for an intervention of ~ -3.5 W/m^2. *Nota bene*: Despite a large error bar, the anthropogenic net at bottom is always positive. This depiction neatly illustrates the dilemma: the terrestrial techniques that are safe do not match the scale of the problem, or take too long to produce a useful effect, while those that do fit the bill tend to be risky or outright dangerous. What is needed is an effective countermeasure, which is both safe and matched in potential scope to the problem, since managing a complicated portfolio of "wedges" incurs cost, too. It would be good if the solution was not too expensive, either, and even better if it paid for itself.

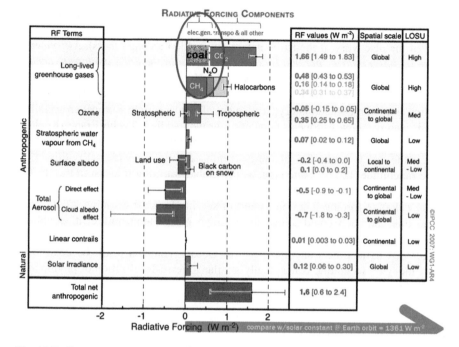

Fig. A3.5 Components of the radiative forcing function in context. Source: IPCC *Fourth Report* (2007), modified by authors

Cloudy or Sunny Days Ahead?

Notwithstanding all the prognostication of the past decades, climate change can still surprise us with its tendency to deliver the unexpected. The U.S. Navy found one not too long ago (2007), when it realized that global patterns of traffic and trade routes will fundamentally change after the Arctic summer ice disappears. Our Navy's primary mission since the beginning of this commercial maritime Republic, like its British predecessor, has been to protect its sea lines of communication (SLOC, the term of art for such routes by those who ponder grand strategy). Its portfolio has just gotten bigger, by one major brief, at a time of diminishing fiscal resources. One hopes that cool heads will prevail in the collegial and increasingly geopolitically significant Arctic Council (Fig. A3.6).

Item: Consider the Arctic Circle. A brief glance at a globe will show the viewer that all the major landmasses, except the two small antipodal continents of Australia and Antarctica, can be closely connecting by running a line around the Arctic littoral. No containership skipper would pass up the opportunity to shave ~5,000 nautical miles and 10 days' steaming time off an east-west journey, say from Yokohama to Rotterdam, by steering north across the top of the world (orange route) rather than threading narrow straits, dangerous seas, and skipping at least five strategic chokepoints (green route) in exchange for just one. Not to mention actual

Fig. A3.6 One unexpected consequence of climate change—faster shipping routes and a new security burden for the US Navy. Image of globe from Google Earth™ (2011), modified by authors. Chokepoints are red "X"s

pirates (bottom of green route). Three of the world's greatest rivers, all north-flowing—the Ob, the Yenisey, and the Lena, each bigger than the Mississippi and each draining a watershed as great as the American Middle West—decant into the Arctic. Each of those three deltas could host a city the size of New Orleans. Major development is likely along the Arctic littoral with the coming of these new transarctic SLOCs. To English-speakers of days gone by it was the fabled Northwest Passage, but to Russian-speakers it is "Северная Морскои Руть" or "Northern Sea Route".

Item: Consider the increasing ubiquity of utility-scale PV here on Earth. Photovoltaic technology has reliably fallen by an average of 1 % per month in nominal dollars for at least the last 20 years. This radical price curve, comparable to the technological revolution embodied in Moore's Law, constitutes an *energy* revolution, and in fact has a name of its own: "Swanson's Law". Using today's installed cost of $3.50 per watt(e) as a baseline, then if the long-term trend holds, we can expect ~$1.50 per watt by the end of the decade. This would definitely attain "grid parity". By 2050, PV power would reach the absurd price of only 4 cents per watt in today's money. (One hesitates to exclaim "too cheap to meter", for that has been said before, with regrettable results.) Then again, having mentioned Moore's Law, one must consider the startling examples of the price-performance curves for computer processing power, speed, memory, connectivity, and especially data storage.

Combine these phenomena with the proliferation of EHV transnational transmission lines. One such scheme would bring solar power from North Africa across the Mediterranean to southern Europe; another proposed project would deliver cheap geothermal electricity to the British Isles from Iceland. Every day, public utilities are gaining operating experience at integrating distributed intermittent renewables such as solar and windpower into the grid. By 2050, Buckminster Fuller's dream of a circumpolar "Dymaxion grid" may already exist, with three major longitudinal transmission spines running the length of the continents (Americas/Oceania, Eurasia, East Asia) from north to south, forever connecting loads anywhere on the nightside with clean sunpower on the dayside. Transmission routes to offtakers on the western and eastern coasts of the Americas, as well as to offtakers in Europe, European Russia, and the Mideast, are depicted in Fig. A3.7 below. (Offtakers in East Asia and Oceania are hidden behind Earth's limb at top.) Dotted portions of the Ring denote submarine sections.

The first step to this sunny future is neither technological nor scientific, but political. Explicitly recognizing carbon-related externalities and setting a correct value on them is a rational path to this bright future, because useful approaches such as the Global Grid would then have value. Remember, wires are pennies-on-the-dollar compared to building more generation. This sets the stage economically for the acceptance and integration of orbital power from satellites built by orbital industry. Orbital power is a stepping stone to the stars thanks to its implicit "anti-carbon" quality.

Material revolutions with profound effects have happened before in human history—the example of aluminum is well known. When first extracted from nature with extreme difficulty, it was considered more valuable than gold and silver at first, so that it was chosen to cap the tip of the Washington Monument instead of the traditional precious metals. Today this rather common but still very energy-dense material is used to hold a single serving of cheap beer or soda pop, after which it is often just tossed away.

There is an even earlier example from much further back in human history—almost before history itself existed—a radical transformation in material cost that had the most profound consequences. We allude to the world-historic transition from the Bronze Age to the Iron Age, which also encompassed a "Dark Age" in between. The interregnum lasted perhaps 400 years, during which many ancient cultures in the Near East disappeared without a trace, leaving few or no clues for us. Ancient iron was worth considerably more than its weight in silver, as shown by its preferred placement in royal tombs. Before the discovery of carburizing iron into steel, worked objects in iron tended to be tiny and sourced from meteoric iron—a rare gift from heaven—and thus were highly prized. After the dramatic cultural transition driven by this technological revolution, the price of iron indexed to silver had dropped by a factor of ~80,000, as attested by records of the time. The 400th root of 80,000 is . . . 1.029! This is the same as the present-day long-term post-WWI consumer price index (CPI)! This completely unexpected and independent result

Fig. A3.7 Notional Circumpolar Grid around Arctic littoral perhaps 37 years from now. Width of interconnects to Global Grid and ring exaggerated for clarity. Image of "Marble" globe from NASA (2013), modified by authors

suggests that we too are living in a remarkable age (but we already knew that). Now turn to Chap. 15 for vision of just how remarkable it is.

References Used in Appendix 3, for Further Reading

J.R. McNeill, "Gigantic Follies? Human Exploration and the Space Age in Long-term Historical Perspective", Chapter 1, in S. Dick (ed.), *Remembering the Space Age*, Proceedings of the 50th Anniversary Conference "Reconsidering Sputnik", at the American Association for the Advancement of Science, Washington, DC, 22 Oct 2007, NASA SP-2008-4703, pp. 8, 2008

H. Zinsser, "Rats, Lice, and History", p.vii, p. 6, Bantam, New York, 1935, 1960

W.H. McNeill, "Plagues and Peoples", p. 4, Anchor/Doubleday, New York, 1976

J. Diamond, "Guns, Germs, and Steel", W.W. Norton, New York, 1997

various editors, "Welcome to the Anthropocene", *Economist*, 26 May 2011

W. Steffen, P. Crutzen, J.R. McNeill, "The Anthropocene: Are Humans Now Overwhelming the Great Forces of Nature?", *Ambio* **36**, no. 8, pp. 614–621, 2007

R. Revelle and H.E. Suess, "Carbon Dioxide Exchange between Atmosphere and Ocean and the Question of an Increase of Atmospheric CO_2 during the Past Decades", *Tellus*, **9**, p. 18, 1957.

R.T. Watson and the Core Writing Team (eds.), "Climate Change 2001: Synthesis Report. A Contribution of Working Groups I, II, and III to the Third Assessment Report of the Intergovernmental Panel on Climate Change", Cambridge University Press, Cambridge, United Kingdom, and New York, 2001; and R.K Pachauri, A. Reisinger, and the Core Writing Team, (eds.), "Climate Change 2007: Synthesis Report. Contribution of Working Groups I, II and III to the Fourth Assessment Report of the Intergovernmental Panel on Climate Change", IPCC, Geneva, Switzerland, 2007

R.M. Bierbaum, J.P. Holdren, M.C. MacCracken, R.H. Moss, P.H. Raven, (eds.), Scientific Expert Group on Climate Change, "Confronting Climate Change: Avoiding the Unmanageable and Managing the Unavoidable", Sigma Xi, Research Triangle Park, NC, 2007

Comte Alexis de Toqueville, Chapter 21 "Future Prospects of the United States", *Democracy in America*, (1835)

D.G. Gimpel, J.A. Falcon, M. Grinnan, P. Grossweiler, R.G. Kennedy, K. Kok, R. Meeker, and S. Unikewicz, Energy Talking Points Series, #1 "Three Prophetic Signs of the End of Oil Exports", American Society of Mechanical Engineers National Energy Committee, 3 Feb 2011

various, United Nations Department of Economic and Social Affairs, Population Division, Population Estimates and Projections Section, "World Population Prospects: The 2012 Revision", http://esa.un.org/wpp/unpp/panel_population.htm, accessed: 28 Jul 2013

N.N. Taleb, "The Black Swan: The Impact of the Highly Improbable", 2nd ed., Random House & Penguin, New York, 2010

C. Marchetti, "On Geoengineering and the CO_2 Problem", International Institute of Applied Systems Analysis, Laxenburg, Austria, 1976

M.I. Hoffert, K. Caldeira, G. Benford, et al., "Advanced Technology Paths to Global Climate Stability: Energy for a Greenhouse Planet," *Science* **298**, 981–987, 2002

Yu. A. Izrael, "Investigations of a Possibility to Reduce the Solar Radiation Flux by a Layer of Artificial Aerosols Aimed at Stabilization of the Global Climate at Its Present-day Level. Results of Field Experiments Carried out in Russia", in Yu. A. Izrael, A.G. Ryaboshapko, and S.A. Gromov, (eds.), *Investigation of Possibilities of Climate Stabilization using New Technologies, Proceedings of Problems of Adaptation to Climate Change PACC-2011*, Moscow, 07-09 Nov 2011, Rosgidromet and Russian Academy of Science ISBN 978-5-904206-12-3, Moscow, 2012, pp. 5–11.

A. Gadian, B. Parkes, and J. Latham, "Marine Cloud Brightening: The Effect on Global Surface Temperatures", in Yu. A. Izrael, *et al*, (eds.), pp. 100–107, *op.cit.*

U. Bathmann, D. Wolf-Gladrow, "Climate Engineering by Carbon Dioxide Removal Techniques: 1. Ocean Alkalinity Enhancement and 2. Ocean Iron Fertilization", in Yu. A. Izrael, *et al*, (eds.), p. 22, *op.cit.*

R.G. Kennedy, K.I. Roy, D.E. Fields, "A New Reflection on Mirrors and Smoke", in Yu. A. Izrael, *et al*, (eds.), pp. 109–120, *op.cit.*

various, GAO, "Climate Change: A Coordinated Strategy Could Focus Federal Geoengineering Research and Inform Governance Efforts", GAO-10-903, September 2010

various, GAO, "Climate engineering: Technical status, future directions, and potential responses", GAO-11-71, July 2011

various, United Nations, "Convention on the Prohibition of Military or Any Other Hostile Use of Environmental Modification Techniques", Adopted by Resolution 31/72 of the United Nations General Assembly on 10 December 1976, http://www.un-documents.net/enmod.htm, accessed: 04 Nov 2012

R. Buckminster Fuller, "Critical Path", St. Martin's Press, 1981

K.I. Roy, R.G. Kennedy, D.E. Fields, "Geoengineering with Solar Sails", Chapter 14 in C.L. Johnson, G.L. Matloff, and C. Bangs, *Paradise Regained: The Regreening of Earth*, 1st ed., Springer, 2010

G. Kopp and J.L. Lean, "A new, lower value of total solar irradiance: Evidence and climate significance", *Geophys. Res. Lett.*, **38**, L01706, 2011. doi:10.1029/2010GL045777

various, "Climate Doctrine of the Russian Federation, unofficial translation", http://eng.kremlin. ru/text/docs/2009/12/223509.shtml, accessed 07 Nov 2011

T.D. Thompson, (ed.), "TRW Space Log: 1957–1991", vol. 27, Thompson Ramo & Wooldridge, Redondo Beach, Calif., 1992

K. Caldeira, G. Bala, and L. Cao, "The Science of Geoengineering", *Annual Review of Earth and Planetary Science*. **41**, pp. 231–256. doi: 10.1146/annurev-earth-042711-105548, 2013

S.H. Salter, G. Sortino, and J. Latham, "Sea-going hardware for the cloud albedo method of reversing global warming", *Philos. Trans. R. Soc. Lond.* **A 366**, pp. 3989–4006. 2008

Military Advisory Board, "National Security and the Threat of Climate Change", Center for Naval Analysis, Alexandria, Virginia, 2007

C. Sagan, "Pale Blue Dot: A Vision of the Human Future in Space", Ballantine Books, New York, 1994

Appendix 4: Mitigating Global Warming Using Space-Based Geoengineering

Robert G Kennedy III, PE; Kenneth I. Roy, PE; Eric Hughes, David E. Fields, Ph.D.

Acronyms

AC, DC	Alternating current, direct current
BAU	Business-as-usual
BoP	Balance of plant
CCS	Carbon capture and sequestration
CDM/CDR/ CDA	Carbon dioxide management/removal/avoidance
CMOS	Complementary metal oxide semiconductor
CPI	Consumer Price Index
EHV	Extremely high voltage
ERO(E)I	Energy-return-on-(energy)-invested
GHG	Greenhouse gas
IOC	Initial operational capability
IPCC	Intergovernmental Panel on Climate Change
JAXA	Japan Aerospace Exploration Agency
Maser	Microwave amplification by stimulated emission of radiation
Molniya	Russian for "lightning", name given to an orbit first used by their space program.
RF	Radio frequency
SRM	Solar radiation management
TRL	Technology readiness level

G. Matloff et al., *Harvesting Space for a Greener Earth*,
DOI 10.1007/978-1-4614-9426-3, © Springer Science+Business Media New York 2014

Introduction to Appendix 4

In Appendix 3, we laid out for comparison the scopes of the challenge of climate change and the global energy system. Their sheer scale dwarfs those of the Manhattan Project and Apollo programs. Despite this, we remind the reader that 1 h of sunshine falling on the bright side of Earth is about the same as the annual primary energy budget of the entire human race today. That's 8766-to-1 ratio...plenty of "headroom" to grow! What if the ordinary cash flows expected in the energy business were harnessed to meet the climate challenge? What if the answer to global warming organically paid for itself in near-real time? What if the overall amount of sunlight hitting Earth could be reduced so that global atmospheric temperatures and weather patterns can return to what we consider normal? What if a large sunshade were built in space to cool off Earth, thus buying time to do the really hard thing: changing our behavior and attitudes?

Leveraging Konstantin Tsiolkovsky's and Fridrikh Tsander's 1924 idea to use mirrors in space for propulsion, Buckminster Fuller's 1940s Dymaxion Grid, Peter Glaser's 1970s study of solar power satellites, and Forward's 1970–1990s concept of "statites", and "Starwisps", in 2001 the authors of this paper proposed to place one or more large photovoltaic/mirrored-solar sail(s) in a sun-centered radiation-levitated non-Keplerian orbit(s) just sunward of the Sun-Earth L1 point (SEL1). This is illustrated in Figs. A4.3, A4.4, A4.5 and A4.6. A earlier version of this concept also appeared in the first edition of the book you are now holding, in 2010.

A Taxonomy of Space-Based Geoengineering

Again, as coined by Marchetti originally in 1976 in the context of terrestrial carbon sequestration, *geoengineering* is the *intentional* application of technology, either ground- or space-based, for the purpose of influencing the properties of a planet on a global scale. So, before describing our "Dyson Dot" concept further, or painting a picture of the "Happy Days Ahead", we recognize that a number of ideas to modify Earth's radiation budget via a space-based approach to geoengineering have been put forth in the community. As a service to that community, we present the space-based half of the taxonomy below, in Fig. A4.1. The terrestrial half, of the taxonomy, masked from the figure below, can be found in the previous chapter. The techniques are separated into the two broad categories of: *solar radiation* management (SRM, left side) and *carbon dioxide* management (CDM, right side). Given the hard vacuum in outer space, CDR does not apply, therefore CDM is comprised solely of clean power generation concepts, which are all subsumed under avoidance (CDA). The techniques are further distinguished by their location in space in order of ascending altitude: LEO-GEO-SEL1.

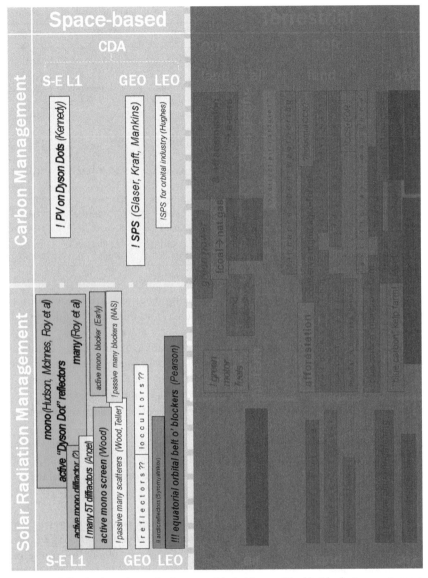

To ensure legibility despite no icons, simple typographical tricks are utilized in this chart as cues to convey six dimensions of data from this mere two-dimensional sheet of paper. **Cost** is not charted. Key:

Effect: tiny 8 pt.=insignificant; small 10 pt.=minor; plain 12 pt.=significant; large 14 pt. **bold=major**; see Table A4.1 next page for numerical values of Effect

Speed/Response: e x p a n d e d   t e x t=very slow or slow (millennia to centuries); plain=multiple decades; *condensed italic=quick (years)*

Risk: ?=unknown; green=none/safe; !yellow=low; !!orange=moderate; !!!red=high/dangerous

Fig. A4.1 A taxonomy of *space-based* geoengineering approaches comparing useful effect/scale, speed, and risk. See summaries

Table A4.1 Terms of assessment and what these evaluations mean numerically or qualitatively

Cost (not shown Fig. A4.1) assessment term	"Negligible"	"Low"	"Modest"	"High"
What it means (US$billions/year)	<1	<10	<100	1000+
Assessment of scale/ effectiveness	"Insignificant"	"Minor"	"Significant"	"Major"
What it means (gigatonnes/year)	<0.3	~1	~3	10+
Approx. forcing function equiv. (W/m²)	<<0.1	0.1	0.3	1+
Speed/response time assessment term	"Very slow"	"Slow"	"Half-fast"	"Quick"
What it means	Millennia	Centuries	Decades	Years
Risk assessment term	"None"	"Low"	"Moderate"	"High"
What does it mean?	Not zero but safe	Think first	Sporty game	Dangerous
Qualifiers	"Unknown"	"Possibly"	"Probably"	*No word*
What they mean	*Unable to assess*	<30 %	>70 %	=Certain

To ensure legibility despite no icons, simple typographical tricks are utilized in this chart as cues to convey six dimensions of data from this mere two-dimensional sheet of paper. **Cost** is not charted. Key:

Effect: tiny 8 pt.=insignificant; small 10 pt.=minor; plain 12 pt.=significant; **large 14 pt. bold=major**; see Table 14.1 next page for numerical values of Effect

Speed/Response: e x p a n d e d t e x t=very slow or slow (millennia to centuries); plain=multiple decades; *condensed italic=quick (years)*

Risk: ?=unknown; green=none/safe; !yellow=low; !!orange=moderate; !!! red=high/dangerous

Taxonomy in words. Summaries of each of the space-based geoengineering techniques follow, listed in the same vertical order as they appear in Fig. A4.1. The more notable items in the list (see Table A4.1) are denoted by **boldface**.

– Spectrum of the Space-Based SRM quadrant:

- *Active occulting mirror, large monolithic or many smaller ones of same aggregate area, in radiation-levitated non-Keplerian orbit(s) at SEL1.* (Roy, Kennedy, Fields, Hughes, 2000–2013) This concept works by casting penumbral shade on target. Orbital mechanics of sunshade described three sections below are common to all concepts based at L1 regardless of optical mode. Power generation co-located on board produces revenue with considerable system mass increase, by ~3×** to ~200–300 megatonnes. Infeasible without cheap launch technology. Common downsides include metastability not true stability, active control to maintain position, unusual feedback-control regime (10-s signal latency, kilosecond natural frequency, effectively no damping). Cost: (possibly extremely) **high**. Effect: **major** (huge). Response: but **quick** (incremental). Risk: **low**.

- *5 trillion active free flying saucer/plate-sized diffractors in cloud around SEL1*. (Angel, 2006) This concept works by diffracting light away from target by ~1°. Some commonalities with orbital mechanics of sunshade described three sections below. Saucer manufacturing and railgun launch technology nonexistent. Specific downside of 5-trillion -free-flyer configuration is local Kessler Syndrome, lack of direct revenue, eventually contaminating cislunar space with migrated debris, numerous dead flyers, and other collision hazards. Common downsides include metastability not true stability, active control to maintain position, 10-s signal latency moot-remote control impossible due to bandwidth for 5 T units, mandatory local control due to high parallelism (5 T) unusual feedback-control regime. Cost: (extremely) **high**. Effect: **major** (huge). Response: **half-fast** (during 5 year deployment). Risk: **moderate**.
- *Large active monolithic screen in orbit at SEL1*. (Wood, 2005) This concept works by screening ~1 % of light from target. Orbital mechanics of sunshade described three sections below are common to all monolithic concepts based at L1 regardless of optical mode. Infeasible without cheap launch technology. Specific downside is lack of direct revenue. Common downsides include metastability not true stability, active control to maintain position, unusual feedback-control regime (10-s signal latency, kilosecond natural frequency, effectively no damping). Cost: (extremely) **high**. Effect: **major** (huge). Response: **quick** (but only upon IOC). Risk: **low**.
- *55,000 active city-size occulting sunshades in school around SEL1*. (NAS, 1992) This concept works by casting many penumbral shadows on target. Orbital mechanics of sunshade described three sections below are common to all monolithic concepts based at L1 regardless of optical mode. System mass hundreds of megatonnes. Specific downside of 55,000-free-flyer configuration is local Kessler Syndrome, lack of direct revenue, eventually contaminating cislunar space with migrated debris, dead occultors, and other collision hazards. Infeasible without cheap launch technology. Common downsides include metastability not true stability, active control to maintain position, unusual feedback-control regime (10-s signal latency). Cost: (extremely) **high**. Effect: **major** (huge). Response: **quick** (incremental). Risk: **low**.
- *Large active monolithic occulting sunshade in orbit at SEL1*. (Early, 1989; Mautner, 1989, Hudson, 1991, McInnes, 2002) This concept works by casting penumbral shade on target. Orbital mechanics of sunshade described three sections below are common to all monolithic concepts based at L1 regardless of optical mode. System masses range ~150–1,000 megatonnes, depending on materials and design details. Infeasible without cheap launch technology. Specific downside is lack of direct revenue. Common downsides include metastability not true stability, active control to maintain position, unusual feedback-control regime (10-s signal latency, kilosecond natural frequency, effectively no damping). Cost: (extremely) **high**. Effect: **major** (huge). Response: **quick** (but only upon IOC). Risk: **low**.

- <u>Passive cloud of scattering elements around SEL1</u>. (Teller, 2002) This concept works by scattering light away from target. Because of passivity, few commonalities with orbital mechanics of sunshade described below. Inefficient mass utilization due to vague fuzzy boundaries. Lack of active position control means eventual displacement by light pressure & solar wind off "sweet spot". Initial system mass 10s of megatonnes, but continual replenishment required. Infeasible without cheap launch technology. Specific downsides include ongoing outlays for replenishment, lack of direct revenue, eventually contaminating cislunar space with collision hazards, short-term irreversibility in case of error. Cost: (continual) **high**. Effect: **major** (huge, but temporary). Response: **quick** (incremental). Risk: **moderate**.
- <u>Cloud of dust around SEL1</u>. This concept works by casting penumbral shade and scattering light away from target. Few commonalities with orbital mechanics of sunshade described three sections below are due to passivity. Inefficient mass utilization due to vague fuzzy boundaries. Lack of active position control means rapid displacement by light pressure & solar wind off "sweet spot". Initial system mass 10s of megatonnes, but continual replenishment required. Infeasible without cheap launch technology. Specific downsides include ongoing outlays for replenishment, lack of direct revenue, eventually contaminating cislunar space with collision hazards, short-term irreversibility in case of error. Cost: (continual) **high**. Effect: **major** (huge but brief). Response: **quick** (incremental). Risk: **moderate**.
- <u>Sunshades (solid occultor or reflective mirror) in GEO or low-inclined Molniya orbits</u>. Satellites in these orbits are eclipsed by earth only a few percent of the time. Duty cycle is reciprocal of redundancy. Hence direct umbral shadowing scheme would have to have very high redundancy (30–100× more shade area than necessary). Molniya orbit must be very low inclination in order for shadow cone to fall on Earth at all; shadow cone from GEO would miss world below most of the time. Many parasols in GEO would be required to do the work of several much closer in at LEO, or one much farther out at SEL1; thus total launch mass, hence cost, would likewise be a large multiple of single parasol. At dusk/dawn and at night, irregular but bright flashes (*viz.*, "Iridium flares") might disrupt the photoperiods of plant life on ground. Valuable orbital slots in GEO already fully subscribed lucrative comsats. Downsides include interference with GEO comsats, astronomy, perceptible shadow, possible harm due to disruption of light-mediated ecological cycles. Cost: (extremely) **high**. Effect: **minor**. Response: **half-fast** (upon IOC). Risk: **moderate**.
- <u>Arctic illumination via lightsail in LEO</u>. Opposite of below. Reflectors in highly inclined LEO to provide or supplement wintertime lighting in polar regions. (Syromyatnikov, *1999*) Not a SRM or CDM approach *per se*, but definitely quasi-geoengineering. Several space experiments attempted over past two decades, inc. *Cosmos-1* sail deployed from Russian launch vehicle (failed), JAXA's IKAROS sailed from Earth to Venus flyby (success). Downsides of industrial/city-scale illuminator include traffic control and

deconflicting valuable low-inclination launch/traffic, interference with astronomy, possible harm due to disruption of photoperiods of plant life and other light-mediated ecological cycles. Cost: **low**. Effect: **minor**. Response: *not applicable*. Risk: **moderate**.

- Sunshades (solid occultors or reflective mirrors) in LEO would have inefficient duty cycle thus moderately high redundancy (~3× more than needed). Each parasol would sweep huge volume and thus be holed by co-orbiting space junk, or collide with hundreds of important satellites. Parasols in darkness for half an orbit, also shadow cone misses world below during max elongation in remaining portion of orbit. Sharp umbral shadow from such large objects would be readily perceptible on the ground. At dusk/dawn and at night, irregular but extremely bright flashes (*viz.*, "Iridium flares") would disrupt the photoperiods of plant life below. Downsides of industrial- or city-scale reflector at twilight include interference with astronomy, possible harm due to disruption of light-mediated ecological cycles. Cost: (very) **high**. Effect: **minor to significant**. Response: **quick**. Risk: **high**.

- *Manmade Saturnian-style rings in equatorial low Earth orbit to eclipse Sun.* (Pearson, 2006) Proposal to deploy either a solid orbital belt, or constellation composed of a very large number of individual satellites, to mimic Saturnian rings. If the orbital sunshade is monolithic (solid), then required material would be a combination of *unobtainium*, *balonium*, and *wishalloy*. If sunshade is not monolithic, then a collision cascade between individual elements is inevitable, mandating the use of uncontrollable "dumb" rocks or dust, not manufactured controllable elements. Hundreds of gigatonnes, possibly a teratonne of material required (about the same as the mass of CO_2 released by humanity since Industrial Revolution), as well as astronomical amounts of money to implement. Highly non-uniform shading effect. Highly intrusive intervention. Long-term stability in LEO unclear. Sword of Damocles: potential energy release to biosphere if parasol reentered at once would be on the order of several teratonnes of TNT. Downsides include severe and acute ecological harm due to seasonally varying but semi-permanent darkness in equatorial belt between Tropics of Cancer and Capricorn, permanent exclusion of all space launch traffic and satellites from equatorial LEO. Cost: (extremely) **high**. Effect: **major** (but catastrophic). Response: **half-fast** (upon IOC, which might take forever, given mass). Risk: (extremely) **high**.

- Power generation concepts in the Space-Based CDM quadrant, CDA section:

  - *Global solar power from Dyson Dot at SEL1 beamed back via millimeter wave.* (Roy, Kennedy, Fields, Hughes, 2000–2013) See entry at top, previous page. 100+ % decarbonization of world electricity sector, possibly entire primary energy supply. Potential annual GHG reduction by midcentury ~50 + Gt/year. Revenue low $trillions per year. High EROI. ~10 TWe net power beamed to Earth from PV-fired maser co-located on board sunshade. ~200–300 megatonnes, or specific system power density = 1 kg per 60 W_e. Atmosphere opaque to primary MMW maser. Infeasible without cheap launch technology. Specific downsides include steering primary maser, active control for sail and relays, unusual feedback-control regime (10-s signal latency, kilosecond natural frequency, effectively no damping). Cost: (possibly extremely) **high**. Effect: **major**. Response: **quick** (incremental). Risk: **low**.

- *City-scale solar-power satellites in GEO beamed back via microwave.* (Glaser, 1968; Kraft, 1979; Mankins, 2008) 5 GW increments. Potential annual GHG reduction by midcentury 20+ Mt/SPS/year. Revenue low $billions per year per SPS. High EROI. 10,000 tons per GW_e, or specific system power density = 1 kg per 100 W_e. Infeasible without cheap launch technology. Cost: (very) **high**. Effect: **major**. Response: **quick** (incremental). Risk: **low**.
- *Industrial-scale solar-power satellites in LEO for orbital industry.* (Hughes, 2013) Not a geoengineering technique per se, but has "anti carbon" avoidance benefits. MW-scale increments. Potential annual GHG reduction by midcentury 4+ Kt/SPS/year. Revenue low $millions per year per small SPS for sales to ground, or ~$billion per year per small SPS for sales to customers in space at estimated current ISS fully-burdened electricity rates of $50–500/kWh. High EROI. 10 tons per MW_e. May be feasible in future even absent cheap launch technology. Cost: (very) **high**. Effect: **insignificant to minor**. Response: *not applicable*. Risk: **moderate**.

More Ruminations About Radiation Forcing Functions (Fig. A4.2)

In Appendix 3, we explained why a "radiative forcing function" of ~ -3.5 W/m^2, was the correct target to shoot for. This is equal to 0.25 % of the 1361 W/m^2 solar constant at Earth orbit, and the same reduction as the *Maunder minimum* (aka *"Little Ice Age"*) when temperatures fell by 2.7 °F (1.5 °C) from the late-sixteenth century to the early nineteenth century. This happens to be the same magnitude, but opposite sign, as the global warming forecast in the IPCC's *Third Report*. So, we present that figure again, but now modified to show a *space-based* geoengineering technique which, unlike any single *terrestrial* geoengineering technique, is a fair match all by itself for the challenge of climate change, not terribly risky and furthermore pays for itself.

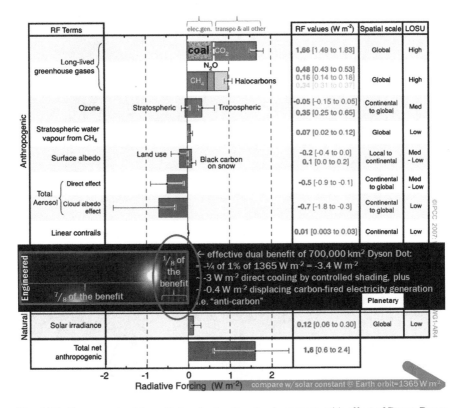

Fig. A4.2 Components of the radiative forcing function in context, with effect of Dyson Dot to scale. Source: IPCC *Fourth Report* (2007), modified by authors

The Dyson Dot

Here is the complete scheme (see here or on color plates) in Figs. A4.3, A4.4, A4.5 and A4.6. Note in Fig. A4.4 that the concept's actual scale is still quite small relative to Earth. The six-petal single-sail concept, as seen from Earth, with relative sizes and distances exaggerated for clarity, is illustrated in Fig. A4.5. (The notional concept is monolithic to simplify analysis.) The opposite view from SEL1 is depicted in Fig. A4.6. Building, placing, and controlling ~700,000 km² of lightsail(s), massing a couple hundred million tonnes, would be the greatest engineering project yet tackled by the human race, but the challenge of worldwide climate change is likewise great. The purpose of this syncretic concept, which we tagged with the moniker "Dyson Dots", is twofold.

1. The 700,000 km² parasol *or multiple lightsails of the same aggregate area* would reduce insolation by an *engineered* forcing function of −3.5 W/m², as

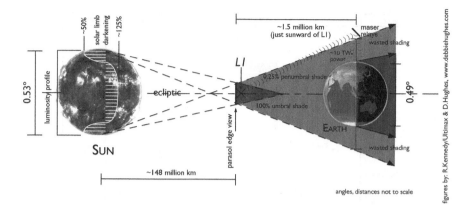

Fig. A4.3 Diagram of single Dyson Dot (edge-on view @ equinox) in non-Keplerian orbit inferior to Sun-Earth L1. Globe of Earth adapted from image courtesy of Wikimedia Commons, itself from NASA. "Earth's City Lights" and "The Blue Marble: Land Surface, Ocean Color and Sea Ice"

Fig. A4.4 Earth, the 900-km Dyson Dot, and a 50-km maser relay (1 of 15), sizes shown in true proportion

shown above. Lowering temperature will reduce the atmosphere's water vapor content, hence the energy available for storms, which should reverse their increasing frequency and severity, likewise limiting the damage from

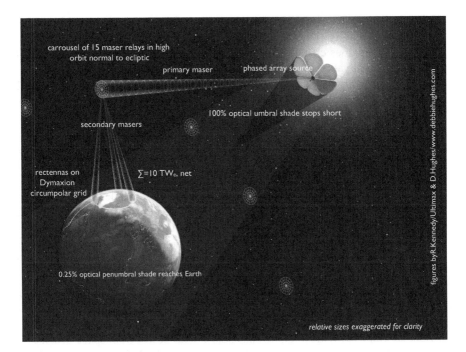

Fig. A4.5 (*Above*). Dyson Dot viewed from Earth. . .

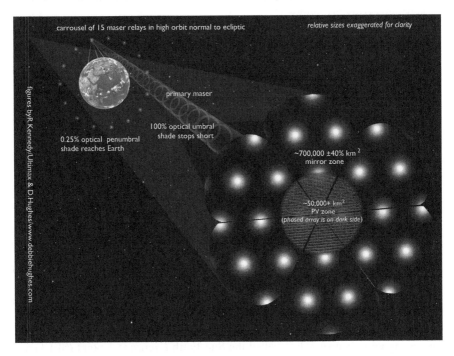

Fig. A4.6 (*Below*). View of Earth and Dot from SEL1

climate change. The nominal value of the sum of these expected casualties over the coming century (integrated from t_{now} to 2100 under the damage curve in Fig. A4.3, previous chapter) if we do nothing is on the close order of US$200 trillion; present value is $20–50 trillion, discounted to 2010 dollars. To maintain its position without continual shipments of expensive rocket fuel, the sail (s) would utilize and manipulate the light pressure imparted by the very photons it diverts from us, an elegant bonus.

2. Partly covered on its sunny side with a ~50,000+ km^2 PV power station, beaming that energy to Earth via phased maser (_m_icrowave _laser_) array, and interfacing with the circumpolar transmission grid described in Chap. 14, the shield could offset ~300 EJ/a (~100 trillion kWh/year, or ~10 TW_e steady-state) of electricity on the ground. This is roughly the same as humanity's global demand for electric power we forecast by 2050 (see Fig. A3.3), which in turn would displace most carbon burners from the terrestrial grid. Notice the "anti-carbon" benefit of a Dyson Dot in Fig. A4.2: the block representing net PV power from space is the same size as the block for coal-fired power at the top of the figure. It is another _engineered_ forcing function, "only" -0.5 W/m^2, but layered for free atop the first one. Revenue from electricity sales of a mere 1 penny per kWh can amortize a trillion dollars of CapEx, each year in nominal dollars, or service the debt in real inflation-adjusted dollars. 3¢/kWh would retire the construction debt outright as well as provide a return This is how solar power from space becomes a powerful method of carbon dioxide _avoidance_ (CDA), i.e., another tool for carbon management analogous to the position of CDR in the terrestrial geoengineering taxonomy.

Avoiding the damage of climate change (insured or not) would provide the long-term hectotrillion-dollar value proposition, while clean energy sales would provide a few trillions annually in actual cash flow to pay for the scheme–there is plenty of potential capital to organically harness without invoking top-down crash programs by _fiat_ or printing money.

How Lightsails Move: Radiation-Levitation and Non-Keplerian Orbits

Gravity keeps our feet on the ground, satellites in orbit overhead, and Earth revolving around the Sun in a Keplerian ellipse. The Sun's (shown as "Primary" in Fig. A4.7 below) immense mass pulls on Earth (depicted as Satellite below)—but for our motion in orbit around it, we would fall in. Earth likewise tugs on the Sun. There exists an orbiting point in space between the two bodies at which an interesting thing happens. It is called the Sun-Earth _Lagrange_-1 point (SEL1) and

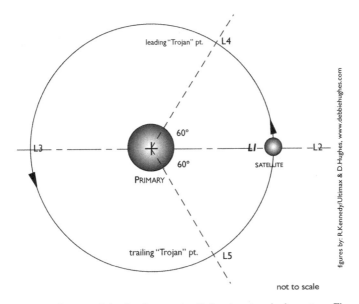

Fig. A4.7 General diagram of the five Lagrangian Points in a two-body system. Figures by: R. Kennedy/Ultimax & D. Hughes, www.debbiehughes.com

is much closer to us than the Sun, due to our much smaller mass. The term honors Italian mathematician Giuseppe-Luigi, Count of Lagrange, who in the eighteenth century worked out the physics to describe the behavior of celestial objects at these points.

Imagine a small object ($m_{Satellite} \gg m_{Object}$, say, 100:1 or better) at SEL1. Because the Earth is on the outside, our gravity accelerates this object exactly opposite to the acceleration caused by the Sun's gravity acting on the inside. From the object's point of view, this interesting position in effect *reduces the mass of the Sun*, allowing the object to have a solar-orbital period *while at L1* exactly equal to that of the planet further out seemingly in violation of Kepler's Laws. Thanks to the vector-sum of these forces, the orbital period of the third body at this particular relative position inferior to us (inside our orbit) is greater (i.e. slower) than it otherwise would be. The SEL1 point revolves in the plane of Earth's orbit, and precisely in phase with it, and thus is *always* on the line-of-sight between the Earth and the Sun. (In fact, while the Satellite *revolves* around the Primary, the set of five points in Fig. A4.7 *rotates* in lockstep with the Satellite about the Primary.) But if the object was a solar sail we would have *another force* to work with—the result of photons hitting and/or reflecting from the sail—*which can be manipulated and controlled by multiple independent physical phenomena (varying reflection, absorption, diffraction) in all six axes of motion* to slow down or speed up the object, thus sinking the object inferior, or lifting it superior, relative to SEL1 itself. Hence the expression "radiation-levitated".

For completeness, we note that the solar wind, composed principally of ionized protons, would have a measurable and nearly random varying effect on the sail also, but on average $1,000\times$ smaller than the light pressure. The resultant electrostatic charging leading to possibly disruptive internal forces will also have to be dealt with. So will the electrodynamic forces generated by motion of charged sails in the geomagnetic fields of Sol and Earth. Luna in her orbit around Earth will impart a measurable but predictable gravitational effect on the sail; so will other celestial bodies but at least an order of magnitude less than Luna's. Earth's orbit itself not a perfect circle, which is a source of more complication for position management—but all these are solvable. But while Luna moves against the background of fixed stars over a month, the SEL1 point (about four times farther away than the Moon) will always be directly between Earth and the Sun. Therefore, an object placed there will always block some of the sunlight that would otherwise hit Earth. This effect is occultation, not an eclipse. Unless the object at SEL1 is thousands of km in diameter, the dark part (*umbra*) of a shadow from there cannot reach Earth—only the dim outer *penumbra* stretches that far. Thanks to geometry and diffraction, the dark spot on the ground during a total eclipse of the Sun by the Moon ("totality") is but a few hundred km across at most, even though the Moon subtends an angle six times as wide as a monolithic Dyson Dot would, and is four times closer to us than a Dot at L1. The reader can appreciate that multiple smaller Dots would be effectively invisible to the unaided eye, satisfying one design goal that the sunshade from a parasol at SEL1 be imperceptible on bare skin. But over the large area of the Earth, it would add up to a significant (and necessary) effect. Sometimes ice ages happen, too, therefore a similar technique could be applied to *raise* the global temperature, because Dyson Dots *reversibly convert* the *solar constant* into a *solar variable*.

Not all shade is equal: see in Fig. A4.3 that a sunshade parked right on the Sun-Earth line-of-sight intercepts a higher quality of light than one parked off-axis. This phenomenon, which varies by a factor of 2.5 or so across the solar disk, is called *solar limb darkening* (or *brightening* for our purposes) and is one of the two components of shading efficiency. This factor improves the effectiveness of a sunshade by $+25\ \%$, compared to the average brightness of the entire solar disk. The schematic also illustrates why no stable sunshade can project exactly the right-sized shadow spot (same diameter as the Earth). Some shade is unavoidably wasted, thus the other component of shading efficiency is maximized at 82 % exactly at L1; all locations inferior to L1 are worse.

Although the Lagrangians are called "points", they are in fact *spatial regions* of *meta*stability (L1, L2, L3) or true stability (L4, L5). These regions are well defined when $m_{Primary} \gg m_{Satellite}$, (say, 100:1 or better). For example, the region of metastability centered around the SEL1 point in Fig. A4.3 is shaped like a sausage lying about 800,000 km along the orbit (i.e., perpendicular to this page). The region's transverse dimensions (i.e., up-and-down, and left-to-right on the page) are approximately 200,000 km each. Objects can orbit the SEL1 Point itself ("halo orbits") as if it were a virtual planet, but at unusually low relative velocities, for the gradient is rather shallow. The dynamics are well understood—for decades, SEL1 has been the favored spot for the world's solar observatories in readily controllable, quasi-periodic, "Lissajous orbits". Thus the SEL1 region contains roughly 30 quadrillion cubic kilometers of space, within an order of the 200-quadrillion km^3 volume of cislunar space in the Earth-Moon system. However, for purposes of shading Earth from excess sunlight, the "sweet spot" is a lot smaller, less than 1 % of that huge SEL1 region.

Why Lightsails?

To shade our planet for the least cost, we need maximum surface for minimum mass. Luckily we have something in our toolkit that is a perfect fit: a solar sail, which propels itself using sunlight. While a photon has no rest mass, it does have momentum. When light hits an object, the photon imparts some of its momentum to that object. The momentum of a photon is very, very small. But in space, without air and wind and in situations where the pull of gravity from Earth or Sun is also small, even sunlight's miniscule push of 5 $\mu N/m^2$ (a millionth of a pound) at Earth orbit, doubled to ~9 $\mu N/m^2$ if the photon is reflected rather than just absorbed. An entire square kilometer would generate but 1–2 lbs of thrust at Earth orbit, but multiplied by enough area and given enough time, even light can be a significant factor in making a spacecraft move. The Pioneer anomaly teaches us that even humble heat radiation must be taken into account over a long enough timeframe. While a detailed design of a radiator is beyond the scope of this chapter, one can imagine that a net disturbance due to the recoil of thermal photons from the Dot during photovoltaic conversion can be minimized by arranging for radially-symmetric thermal emission normal to the axis of the Dot. The force on the 700,000 km^2 Dot would equal the weight of a jumbo jet. A typical solar sail needs a mass:area ratio (σ) of only 10–20 g per square meter to be useful. But parasols will have major differences compared to "traditional" solar sails used to haul cargo around the solar system. Solar sails for geoengineering can be over ten times as heavy if necessary. They will not have a payload, they *are* the payload. Since they will not be boosting freight, we can make them any size we want based on ease of fabrication

To distinguish these rather specialized lightsails from ones intended for propelling spacecraft, the authors coined the term "Dyson Dot". The moniker is a deliberate allusion to the eponymous Dyson Sphere conjectured by Freeman Dyson (1960) as a system of orbiting space facilities that completely encompasses a star, built by an advanced extraterrestrial race to capture the entire matter/energy output of their sun. In contrast, our Dyson Dot would manipulate only a tiny fraction of a star's light shining on one of its planets (ours). With photovoltaics on their bright side, these giant sails could easily make electricity vastly in excess of their own infinitesimal requirements for housekeeping.

Given the same light pressure, a lighter-mass-density sail will experience greater acceleration than a heavier one. The brighter or less dense the mirror, the less it costs, but the further inside SEL1 it must go. The further the sunshade cruises inside of SEL1, the more shade will be wasted going uselessly past Earth's limb. Paradoxically light weight increases overall mass and cost, and the less efficient the total project gets. The darker and heavier it is, the closer to L1 it can go. Any real solution will be an optimization-tradeoff among multiple values and constraints. For cooling Earth, it does not matter how the Dyson Dots stop solar radiation just as long as it does not get here. Sail mass increases steeply above $\sigma = 50$ g/m^2.

Also accounting for both the solar limb brightening on the S-E line-of-sight which improves effectiveness, it appears at this writing that the optimum parasol parameters to achieve the -3.5 W/m^2 goal are: mass density 53 g/m^2, position ~2 million km from Earth, a bit under 700,000 km^2 of sunshade area.

For purely radiation management purposes, we do not care if we have a 700,000 small sails a kilometer square each or one great monolithic sail the size of Texas. It is the optical properties of the aggregate which are critical. Individuals may vary a lot depending on the particulars. They do not have to be totally reflective mirrors, or totally absorptive blackbodies. Different parts may be mirrored, or black, or diffractive, even transparent, or some combination. The Japanese IKAROS lightsail has all these characteristics. However, passive objects would tend to quickly get displaced into useless positions off the light axis, or pushed out of SEL1 altogether, by light pressure and the solar wind. That is the trouble with the "cloud of dust" techniques—they would not last to pay back to investment.

Photovoltaic materials must have a minimum thickness in order to absorb light. For silicon cells, this is about 30 µm per absorption factor of e in the visible band. 3 factors of e yields about 95 % capture, so a 100 µm thickness of silicon plus a foil backplane would be needed. This indicates a mass density of ~300 g/m^2, significantly denser than a typical lightsail. Again, however, more mass allows the Dot to stay closer to SEL1, improving several parameters. In any case, the PV portion only occupies a fractional patch in the center.

The weapons potential of the primary dekaterawatt transmitter is intentionally precluded by physical design. (Dyson Dots are intended to *solve* a major problem, not create another one.) We choose a 5-mm wavelength (i.e., 60 GHz frequency, presupposing the ability to do power electronics at 60 GHz) because it is strongly absorbed by the O$_2$ in the Earth's atmosphere (1/e in the top 5 %). The extremely high frequency also provides a very tight beam, which is useful given the "long throw" (>2 M km) from SEL1 to Earth.

Therefore secondary relays will be necessary for downlinking to rectennas on Earth via a window that the atmosphere will pass. We choose a much lower frequency (2–3 GHz), and longer wavelength (10–15 cm), for this purpose. (2.45 GHz was selected to minimize ionospheric heating by Glaser and in the NASA studies 1968–1980.) The longer wavelength is acceptable also because beamspread is not a problem for the relatively "short throw" from the relay to the Arctic Circle where the rectennas are. We propose that these relays orbit high, perhaps 10,000-km to avoid the worst of the Van Allen Belts, somewhat elliptical for good dwelling behavior. They revolve in a single plane highly-inclined at 66.5° to Earth's axis, *hence normal to the ecliptic*. Being perpendicular to the ecliptic means they would never transmit to a receiver lower than the Arctic Circle. A single

Table A4.2 Mass partition of the Dyson Dot system

Major blocks of the Dyson Dot system	Mass [Mt], A $\approx$ 690,000 km^2
Lightsail only, for $\sigma = $ **53** g/m^2	**36.5**
Photovoltaics @ 300 g/m^2	*~15.4*
Maser array @ 1 kg/kW, $\eta = 80$ %	*~16.3*
Allow +1 kg/kW for 15 maser relays, $\eta = 80$ %	*~97.9*
Fractal structural web & balance of plant	*~46.9*
Total system mass in space [megatonnes]	*~213.*

ring constellation of 15 birds 24° apart would ensure that a minimum of 2 relays are always over the Arctic Circle at any time. We assume these relays will function as thermally-limited blackbodies, again at 80 % link efficiency. Each relay hands off duty to the next as it revolves into position like a carrousel, much like Chief Designer Korolev's *Molniya* comsats were intended to handle long-haul communications across the far north of the Soviet Union in the 1950s, saving money over expensive land lines. Each secondary beam is steered to a fixed individual receiving antenna on the ground. The rectennas' optimal disposition is strung along the circumpolar hub (described at the end of Appendix 3) like beads on a necklace. Despite their high latitude, the rectennas are always in sight of, and illuminated with, RF energy beamed in a bank shot from the Dot. Simultaneously they are always connected to the Global Grid on the Arctic Circle as described in the previous chapter. We have not sized the relays at this stage, but the 5-GW$_e$ microwave transmitters in the Glaser and NASA studies were 1 km in diameter, which suggests that our relays would be ~50 km across if they used 1970s technology. The aperture of the individual transmitters would be far smaller than the relays themselves, on the close order of 100 m, in order to physically limit the averaged received microwave power down below to a biologically safe 100 W/m^2. On the ground, this means 10-km diameter rectennas to accept industrially-useful amounts of power. In actuality, spanning less than 0.2 % of the breadth of the Arctic Circle, the entire ring would be virtually indistinguishable from space. Mounted above the ground on stanchions, the open mesh-like rectenna array of half dipoles passes 80 % of sunlight, thus does not exclude other simultaneous land uses, such as agriculture and forestry, similar to the way tree plantations are grown under transmission lines, or wind farms and crop farms share space today.

The sail itself is only about one-third of the system's mass. The gigantic 213-Mt number (see Table A4.2 below) is driven mostly by our financial requirement to generate revenue, which means generating *and then delivering* power. Another one-sixth is composed of 15 megatons of PV, driving a 16-megaton, 80 %-efficient long-range primary maser with 16 TW of input power. The bulk of the mass number, ***100 megatons***, is due to a ring of 15 secondary short-range maser relay satellites, also 80 %-efficient, required to deliver a net 10 TW$_e$ to the paying customers on Earth. Just the high polar relays are half the system! We assume that masers can be built to continuously handle 1kW$_{input}$ per 1 kg. That kind of power density sounds aggressive (it is 10 times the density of Glaser's 1968 SPS concept), but consider: Unlike lasers, masers have languished in the technological dark since the discovery of stimulated emission at the dawn of the Space Age. Fortunately, recent news suggests this is changing. According to the authors' personal experience in the late 1970s, a laboratory-grade helium-neon laser was a metal box half a meter long, massed a couple of kilograms, cost ~US$1000, and required a 100-W 110VAC power supply which plugged into the wall, just to generate a few *milli*watts of coherent light. Nowadays, a laser of equal power (and more colors) is but a centimeter in size, a few grams in mass, and runs on batteries—a throwaway consumer toy available to anyone for a few dollars. These chained improvements for laser technology in less than four decades span roughly eight orders of magnitude. We can expect great improvements in the specific power and efficiency of masers if history is any guide. NASA's SPS study in 1979 was predicated on a DC-to-RF link efficiency of 60 %, but 80 % or better with solid-state components is credible today. More efficient transmission methods would have cascading benefits for the entire system.

The Kessler Syndrome—cascading collisions which are already a problem in LEO due to "space junk"—will be a major consideration in the systems engineering. It is a nonlinear function of the number of objects, but this is ameliorated by the vast extent of the SEL1 region and its low velocity gradients with respect to the Point itself, compared to the crowded hypervelocity environment in LEO. To avoid the operational expense and logistical headache of continual replenishment due either to collisions or simply being passive objects, our sails must be "smart": sensors, controls and onboard intelligence. To do their job and remain precisely in their assigned places, they must vary the thrust resulting from solar radiation to counter forces that would pull them off-station. IKAROS's experience suggests that just a fraction of a percent is needed for "solar trim tabs" to control attitude. Because many thousands or possibly millions of solar sails will be cruising in some proximity to one another (much like a giant school of fish) they must be social, i.e., their sensors will observe their neighbors as well as the Primary and the Satellite, and they will maneuver to avoid crashes or other conflicts such as cutting off a neighbor's light. They must have the means to receive and execute instructions from their builders and operators. We will want them to readily move out of position (yet still generate valuable power) if we succeed too much, or because sometimes ice ages happen too. Once in space, Dyson Dots could be applied to *raise* the global temperature, because they *convert* the *solar constant* into a *solar variable*. And they do it *reversibly*, the top design goal for *any* geoengineering scheme.

While this control problem also scales as N-squared, on the other hand, this "school of fish approach"—a larger number of smaller sunshades—captures the value of the learning curve as well as providing incremental benefits of shading for incremental cost without waiting for full system to come online. "Pay a little, build a little, get a little, learn a lot." Incremental engineering tends to reduce overall technical and economic risk. "Continuous production, continuous improvement, equal cheap and high quality good." This outlook and approach is likely to appeal to investors and others who willingly underwrite the scheme.

A single proton hit can flip a bit, or even kill a logic gate outright, due to the microscopic feature size and radiation intolerance of modern integrated circuits based on CMOS technology, compared to the robust vacuum tubes of days gone by. Fortunately, a viable candidate for service under the harsh conditions of space has emerged: "voting micro vacuum tubes" with a technology readiness level (TRL) = 4 have been demonstrated in the lab. These are fundamentally redundant and robust in high temperatures and radiation fields, and provide a computational mass density at least equal to 1970s integrated circuit technology. We remind the reader that the venerable Space Shuttle flew for its entire life with 1970s computing technology.

Rather than a one-shot cargo mission, Dots must be built for longevity. (Field results now suggest that polycrystalline silicon PV technology on Earth may last as long as a century in service.) Dots must endure the harsh conditions of interplanetary space, withstand intense radiation for decades and continual assaults by the solar wind (or even the occasional solar storm), and tolerate occasional punctures by micrometeoroids.

The radiation-*cum*-longevity design requirement would exclude the common polymers such as Mylar™ that we know, but perhaps an exotic composite based on carbon, fluorine, and silicone can be found. Iron is certainly strong enough and common enough in the Solar System, but in this application there will be a strong premium on strength:mass ratio. Candidate metals which are both light and strong include magnesium, aluminum, titanium, lithium and beryllium, in order of mass-abundance in the solar system. But the latter two are as rare as niobium and uranium, respectively, plus the last one is very hazardous to work with. Speaking of "rare", gold could do the job! Regrettably the required quantity of gold leaf for a lightsail the size of Texas—even if hammered into foil just 200 Ångstroms thick—is a quarter-million tons, exceeding what has been mined in all of history so far.

A fractal web in the backplane for ripstop and reinforcement, combined with built-up structural techniques that Zeppelin engineers would recognize, plus "balance of plant" (BoP, means everything else), brings the system mass—sail, structure, BoP, PV, masers, and 15 relays (see detail in Table A4.2)—to a grand total of 213 million tonnes.

213 megatonnes sounds like a lot, but for perspective bear in mind that the world burns about four *billion* tonnes of coal, and another six *billion* tonnes of oil and gas, *every* year. One supertanker of the many hauling crude oil around the world's oceans weighs about half a million tons fully loaded; and 200 Mt is about two week's worth of production.) Speaking of oil, the Sun-source-to-Earth-sink chained efficiency for the entire Dyson Dot system is a respectable 14.78 %, about the same as the so-called "well-to-wheel" efficiency of a typical fossil-fueled automobile! (*Nota bene*: the gas buggy's reported efficiency does not account for the solar energy which made the petroleum in the first place, over millions of years at an extremely low primary efficiency of ~0.3 %.

In the grand scheme of things, 213 megatonnes is not so much. For example, it is the mass of a small stony-iron asteroid less than 400 m across–a class of rock so

Fig. A4.8 Another simple syllogism: one latent positive-feedback loop waiting to take off in space

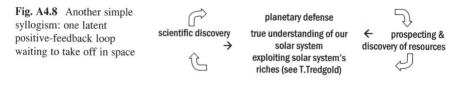

scientific discovery →

planetary defense
true understanding of our
solar system
exploiting solar system's
riches (see T.Tredgold)

← prospecting &
discovery of resources

minor that we have not seriously looked for it yet. But when we do find that rock, we are sure to need it! Ten years of construction means assembled 60,000 tons (mass of a skyscraper) of *finished* Dot stuff per day, which in turn means handling maybe 100 times that much raw material, possibly more. Just the act of building the Dots provides a development path of important mutually-reinforcing capabilities (Fig. A4.8).

Asteroids, of the metallic and carbonaceous chondrite variety, are the most source of building materials for the Dots, based on our limited understanding today. Certainly it is far more cost-effective than simply hauling hectomegatonnes of mass out of Earth's deep gravity well with inadequate space launch technology, or even off the Luna or Mars. Asteroids are more likely to contain the necessities, especially volatiles and primordial unbound metals, than the blasted regolith blanketing our Moon. The old space geology joke goes, if we found concrete on the Moon, we would mine it for the water. In the cosmos, "common" aluminum is actually an order of magnitude rarer than magnesium. Al is over-represented in Earth's crust due to the refractory yet lightweight compounds it forms with oxygen, hence they survived and floated with the leftovers during Earth's accretion and differentiation. Comparing its abundance relative to copper, one can see that Al would be material of choice for power distribution up there, just as it is down here on *terra firma*. Titanium is even rarer than aluminum and strongly binds with oxygen to form refractory compounds that are also difficult to crack. Titanium is stronger but denser, harder and much less conductive and workable than aluminum, while magnesium is lighter, softer and more ductile (i.e. fabricable). Perhaps an exotic alloy possible only in space is the answer—microgravity metallurgy *in vacuo* is *terra incognita*. But thanks to nature's preference for even-numbered atomic numbers in elemental occurrences (see Fig. A4.9 below), a mostly-magnesium-alloy may turn out to be the preferred choice in space for a light, strong, durable engineering material rather than the more well-known aluminum. Notice how magnesium is as common as iron, or as the silicon in sand or glass.

Whatever sail material is eventually selected, if it is not available in high purity then at least it must be common, in order to minimize the mass-processing requirement on the front end of mining and extraction. In essence, the builders must *impedance match* the Dot's use of elements with their abundance as found in asteroidal material in order to minimize the amount of waste mass being hauled around and not utilized, which acts as a cost multiplier on the mining cost. Stony asteroids, the second most common types after chondrites, are made of silicates of Mg, Fe, and Al. Another obvious class of engineering materials to develop is ceramics. Carbon, nitrogen, and oxygen will comprise roughly half of the total mass that is found in asteroids that we know about—carbides, nitrides, and oxides

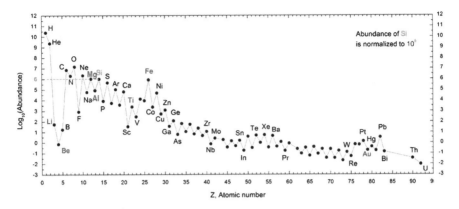

Fig. A4.9 Odd/even elemental abundances in the solar system, on a logarithmic scale. Source: Wikimedia Commons, cross-checked with *CRC Handbook of Chemistry and Physics*

of the transition metals are all ceramics. SiC, for example, is an excellent high-temperature semiconductor, so it could become an important piece in the technology chain for power conversion in space. Chemical engineering methods in space, such as non-redox techniques for the extraction of metals in microgravity are also virtually unexplored territory. The relevant literature is about as featureless as the medium it would operate in. Simply converting "unknown unknowns" to "known unknowns" would be a valuable contribution; so would developing a taxonomy of techniques and a grammar of equipment. Filling in this gap in our know-how constitutes a ***rich*** field for future scholarship and engineering R&D.

Since the end-product is a semi-autonomous robotic vessel with an endless supply of fuel, we can locate the factories where the mineral resources are, not the end-users/customers/buyers. Once the Dot slides down the ways of the photonic shipyard, so to speak, gets checked out and commissioned, it can deliver itself anywhere. We trust that they're smart enough to get to their assigned spot. A team of bright young motivated Japanese people demonstrated this nicely, on a remarkably small budget (~$20 M) in an incredibly short time (2-½ years) with their IKAROS interplanetary lightsail.

Once the main objectives of the first Dyson Dot(s) have been met, and a nascent space industry is in place, we suppose that many other large-scale uses will be found offworld for Dyson Dots and the energy they produce, first at cislunar ranges, then perhaps longer. Can a Dyson Dot catch the solar wind and accumulate a worthwhile amount of ^{3}He over a lifetime of service catching rays? Another application could be direct reduction of asteroidal metals at Inner System planetary ranges. Then even more ambitious uses could be found, perhaps including the boosting first-generation probes to velocities and ranges suitable for interstellar exploration.

In principle, geoengineering methods could be applied to any other planet, anywhere. The L1 regions (not to mention their Trojan points) for the Sun-Jupiter or Sun-Saturn ordered pairs are much larger than the SEL1—in fact, far greater extent than the gas giants or even Sol himself. These outer regions could host mirror-sails in large numbers—*vast*. Given the Copernican Principle, none of these considerations is unique to our home system. Physical laws are the same for everybody. It is reasonable to suppose that intelligent beings elsewhere may apply techniques we would recognize. If a race were interested in terraforming a planet, solar sails would be a useful technique. Dyson Dots as described here can *convert* a *solar constant* into a *solar variable*, and to modify the color of light hitting the target planet. They can catch light from their star for energy resources, and reap their stellar wind for material ones.

Although we may find the project too daunting right now, other beings in other solar systems may have done it already. Perhaps researchers at the Search for Extraterrestrial Intelligence (SETI) Institute should look for the occasional flash from distant Dyson Dots, or leakage of microwave power from the transmission system. Dots are designed to reflect a lot of energy, perhaps even alter the color of sunlight. A school of mirrors at the L1 point pairing a jovian planet and central star in some other system would have much greater diameter than the gas giant or star itself. This bright source would appear to wax/wane by many magnitudes, yet be synchronized with the orbital period. The reflection would display rhythm but due to the non-Keplerian nature of sails' orbits, the observed motion would seem anomalous somehow. If the instruments doing the observing were sensitive enough, this space oddity might be visible at interstellar range to outsiders like us. They may not want to talk to us, but it would somehow be comforting to know that other beings are making themselves comfortable on distant worlds.

Conclusion: The Killer App

Readers of a certain age may recall that a "killer app" is not a merely useful piece of software, rather it is an application which is so democratically useful as to cause a sea change in demand, hence proliferation, which then drives the cost of the technology downward in a strong positive-feedback loop. Such transitions tend to be rapid and quasi-ecological, with unanticipated effects. Two examples from just the history of computing include: the early spreadsheet called Lotus 1-2-3 created widespread demand for desktop computers which spawned the personal computing revolution; an early computer game called "Myst" created widespread demand for CD-ROM media and the boom in cheap digital storage technology.

Chemical rocket technology is completely inadequate in both performance and scale to implement Dots. At a current launch cost of $10,000/kg, a 200-megatonne Dot above would cost ~$2000 trillion ($2 \times 10^{15}$ dollars, or gross planetary product of all goods and services for half a century) just for the ride into LEO. Clearly, *reductio ad absurdum*. (However, the marginal cost in energy to climb further up the potential well from GEO to L1 is practically negligible.) Compare also the mass of a Dot to the sum of payloads permanently launched into space by about 5,000 rockets since October 4, 1957: a mere 5 *kilo*tonnes, or an average of ton a shot. The

Fig. A4.10 The Killer App, and a roadmap to the stars

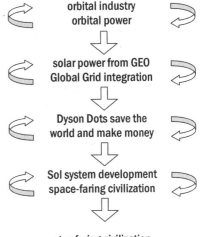

orbital industry
orbital power

solar power from GEO
Global Grid integration

Dyson Dots save the
world and make money

Sol system development
space-faring civilization

star-faring civilization

scale of the energy problem likewise dwarfs the Apollo project. Only a truly spacefaring civilization with cheap access to orbit, offworld materials, and using energy on an unprecedented scale could execute this project. However, proposed solutions to global warming—including business-as-usual (BAU) which is always an option in human affairs—are also very expensive or painful. Perhaps becoming an advanced space-faring civilization would be the cheaper alternative after all: Like the highly-mass-efficient leaves in an ecosystem which turn sunlight into food, PV lightsails capture and convert a great deal of energy at a very low mass density better than any other non-nuclear energy technology we possess. The Dot's power density of ~30 kilotonnes per GW_e, measured at the end user, compares favorably with nuclear electricity generation. Can we deploy Dyson Dots in time to make a difference in global warming here? Certainly not without far cheaper and more reliable access to space, nor without the national or international political will to do great things. It is clear that building and using Dyson Dots would be the biggest single engineering project yet attempted by the human race, but it need not be a top-down *fiat* effort. The project could be organically self-funding because known or expected cash flows for energy and infrastructure are about equal to the task, while the cost of doing nothing in the face of climate change would be much greater. Building such megastructures with offworld materials ("living off the land in space") would create a set of mutually reinforcing capabilities, each of them valuable, perhaps even indispensable—the *sine qua non*—to a spacefaring civilization. As the Englishman Thomas Tredgold defined it almost two centuries ago for the then-nascent British Institution of Civil Engineers, "*engineering is . . . the art of directing the great sources of power in Nature for the use and convenience of man*".

Solving the climate change problem here on Earth may be the "killer app" to proliferate Dyson Dots and bootstrap the human race *ad astra* to a new plane of existence (Fig. A4.10).

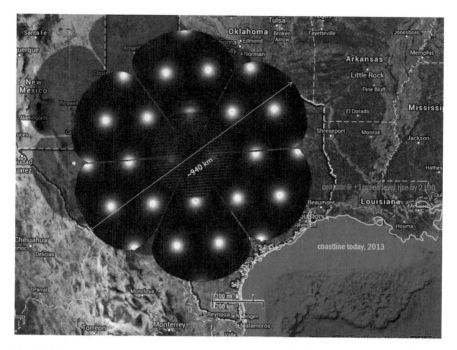

Fig. A4.11 The Dyson Dot is the size of a small country, or Texas. Note area of flooded real estate between present day coastline and 2100 sea level contour under BAU scenario. How would sum of all drowned land compared to the Dot's area? Background from Google Maps™ (2012); contours from USGS

"Happy Days Ahead?" (Fig. A4.11)

So here you are in the future, a peripatetic roughneck, willing to travel to do the hard work you like, and also because you were tired of getting your feet wet. Those space factories got built in cislunar space with captured rocks, and some are cranking away far deeper in space, where the really good stuff is. (You know all about moving to where the work is.) The space fabs have been churning out Dyson Dots for 10 years, which upon commissioning and check-out, dutifully hie themselves Sunward to the Lagrange-1 point. Occasionally one of them takes a flyer (so to speak) to a different destination on a more interesting mission. A robust program of asteroid detection, characterization, prospecting, capture/retrieval, mining and extraction, in support of building the Dots turns out to have a lot cost less than launching a complete Dot off the ground *ab initio*. (As if! And your feet would still be wet...like up to your neck!) The program dug up loads of information about the distribution and behavior of asteroids (and comets), that was directly applicable to planetary defense and a trove for science to mine. We dodged a few bullets, though not all of them. (You don't know what you don't know. People in this part of the world are funny about that kind of thing in particular—they're the only ones it seems to happen to. Well, that's what you get for being the biggest target.) And *vice versa*: the search/survey program motivated by defense or science was useful to the prospectors and miners. Notwithstanding motives high and low, more eyes in space have been an absolute good for everyone, and no doubt will continue to yield serendipitous scientific discoveries.

So today you read on your widget that the scientists and engineers running this crazy scheme announced that today's the day we will reach our goal. We will finally have enough mirrors at the proper point to reduce peak global temperatures to a level not seen in decades. . .and take the edge off everything else that came with that. (It hasn't been this calm in years, either.)

Naturally, a world holiday has been declared! You look through your window of your mobile at the sun lollygagging not far above the horizon as it does during the white nights up here. (A roustabout can get used to anything, and besides, the endless days mean more time to work, which means more time to make money. It's been a little strange slinging cable instead of drill pipe, to move juice instead of crude, and the technical slang is completely different. But some things never change, the pay's all right—better than all right—and the company ain't bad either.)

There's a better view outside, but you don a windbreaker before going, now that the top of the world is cooling off again. (That aloha shirt just looks silly now in this sudden chill.) On the way out, you grab those garish safety glasses the barmaid last "night" handed out for the occasion. (She said they went with your shirt.) You put them on, squint, tap them to focus, and see. . .what? Nothing, really. You can't detect a Dyson Dot as far away as the SEL1 point even if it is the size of your old home, not against the backdrop of the Sun itself, because the dark umbra never got to you. Nor do you feel the quarter-percent shadow on your skin, any more than you feel one from that bird flying overhead. (Nowhere to go but south.) Using special devices designed to image sunspots, you might see a strange cluster of black dots in the center of the sun if you were willing to make the effort. To most people going about their everyday affairs on this world, however, the Dots are invisible, like all good infrastructure.

At the start of the Project, the farmers in this future pointed out that the quarter-percent reduction in sunlight would mean a quarter-percent drop in their crop yield, and they were already under pressure as it was. (Or under water.) Yet these folks, no less ready to whine about the weather than they were back in the twentieth century (AD or BC, probably no difference), have been assured that no such shortfall would happen. It turns out that plants are not very efficient at using the natural sunlight landing on them. Chlorophyll uses only narrow bands of light in the blue-purple and orange-red regions of the spectrum. Paradoxically they don't use green at all which is right near the peak (most abundant energy) of the solar spectrum. Green light is bounced away, which is why green plants look green to your eyes. (What a waste! Somebody should do something—how about "black chlorophyll"?) So a few of the Dyson Dots have been modified to assist dirtside farming by shifting a tiny bit of the stupendous energy they collect to the particular colors chlorophyll uses. This supplemental light is beamed to Earth with weak lasers from special Dots. There is even talk of boosting these frequencies above natural levels, but only for the

Earth's agricultural regions. You don't expect to see that at this latitude, though. For one thing, the shorter wavelengths wouldn't make it through the soup on the horizon. For another, the extra tinge of purple would make the sunset just too weird to cope with, even to jaded late-twenty-first century eyes like yours.

Best of all to most voting stockholders, including you, is the recurring dividend which the Dyson Dot Consortium has been putting in your pocket, and every other open pocket on Earth. (Even the "whingin' poms" who don't care for all this newfangled geoengineering stuff still cash the check downloaded to their wallets.) The royalty off the power flowing south from the Arctic Grid saved your Russkii neighbors, too, when their mainstay oil exporting went south also, after the big scare earlier in the century. The insurance is cheap up here 'cuz bad things have stopped happening, and the people are easy enough to get along with. Maybe you can relax for a change. Some things about this boom town even remind you of home. Inspired by that thought, you pat your wallet happily on the way to find a good party...maybe that new Russian-Cajun fusion place that opened just on down the shore, *The Balalaika and the Banjo*.

References Used in Appendix 4, for Further Reading

C. Marchetti, "On Geoengineering and the CO_2 Problem", International Institute of Applied Systems Analysis, Laxenburg, Austria, 1976

M.I. Hoffert, K. Caldeira, G. Benford, et al., "Advanced Technology Paths to Global Climate Stability: Energy for a Greenhouse Planet," *Science* **298**, 981–987, 2002

R.G. Kennedy, K.I. Roy, D.E. Fields, "A New Reflection on Mirrors and Smoke", in Yu. A. Izrael, *et al*, (eds.), pp. 109–120, *op.cit.*

various, Government Accountability Office, "Climate Change: A Coordinated Strategy Could Focus Federal Geoengineering Research and Inform Governance Efforts", GAO-10-903, September 2010

various, GAO, "Climate engineering: Technical status, future directions, and potential responses", GAO-11-71, July 2011

F.A. Tsander, translated as "From a Scientific Heritage", NASA Technical Translation TTF-541, NASA JPL, 1969

K.E. Tsiolkovsky, "Extension of Man into Outer Space", 1921. See also K.E. Tsiolkovski, "Symposium on Jet Propulsion #2", [United Scientific and Technical Presses], Moscow, 1936

R.Buckminster Fuller, "Critical Path", St. Martin's Press, 1981

P. Glaser, "Solar Power Satellites", Arthur D. Little Inc., 1968

C.C. Kraft, "The Solar Power Satellite Concept: The Past Decade and the Next Decade", The Von Karman Lecture presented at the Fifteenth Annual Meeting of the American Institute of Aeronautics and Astronautics, NASA-JSC-#14898, US GPO 1979-673-662, pp. 6–8, 1979.

R.L. Forward, "Statite: Spacecraft That Utilizes Light Pressure and Method of Use", US PTO, patent # US5183225 (A), issued 02 Feb 1993

K.I. Roy, "Solar Sails: An Answer to Global Warming?", CP552, M.S. El-Genk (ed.), *Space Technology and Applications International Forum-2001*, American Institute of Physics, New York, 2001

K.I. Roy, and R.G. Kennedy, "Mirrors & Smoke: Ameliorating Climate Change with Giant Solar Sails", in B. Sterling (ed.), *Whole Earth Review*, p. 70, Summer 2001

B. Fagan, "The Little Ice Age: How Climate Made History: 1300–1850", Basic Books, 2000

various, "Statistical Abstract of the United States", 2011 edition, US Dept. of Commerce, Bureau of the Census, USGPO, 2011

J.L. Lagrange, "Essai sur le Problème des Trois Corps," *Oeuvres de Lagrange*, vol. 6, pp. 229–332, 1772

C.R. McInnes, "Solar Sailing: Technology, Dynamics and Mission Applications", Springer-Praxis, Chichester, UK, 1999

C. Wiley, "Clipper Ships of Space", *Astounding Science Fiction*, May 1951, p. 135.

L. Friedman, "Starsailing: Solar Sails and Interstellar Travel", Wiley, 1988, pp. 9–10

G. Vulpetti, C.L. Johnson, and G.L. Matloff, "Solar Sails: A Novel Approach to Interplanetary Travel", Springer-Praxis, 2008

K.I. Roy, R.G. Kennedy, D.E. Fields, "Geoengineering with Solar Sails", Chapter 14 in C.L. Johnson, G.L. Matloff, and C. Bangs, *Paradise Regained: The Regreening of Earth*, 1st ed., Springer, 2010

J.T. Early, "Space-Based Solar Shield to Offset Greenhouse Effect", *JBIS*, **42**, pp. 567–69, 1989

C.R. McInnes, "Minimum Mass Solar Shield for Terrestrial Climate Control", *JBIS*, **55**, pp. 307–11, 2002

National Academy of Sciences Panel on Policy Implications of Greenhouse Warming, "Policy Implications of Greenhouse Warming: Mitigation, Adaptation, and the Science Base", National Academy Press, Washington, D.C., 1992

R. Angel, "Feasibility of cooling the Earth with a cloud of small spacecraft near the inner Lagrange point (L1)", *Proceedings of the National Academy of Sciences*, **103**, no. 46, pp. 17184–17189, November 14, 2006

E. Teller, R. Hyde, L. Wood, "Active Climate Stabilization: Practical Physics-Based Approaches to Prevention of Climate Change", UCRL-JC-148012, National Academy of Engineering Symposium, Washington, D.C., 23–24 Apr 2002

D. J. Kessler and B. G. Cour-Palais, "Collision Frequency of Artificial Satellites: The Creation of a Debris Belt", *Journal of Geophysical Research*, **83**, pp. 2637–2646, 1978

H. Monohara, M. Mojorradi, R. Toda, R. Lin, A. Liao, "'Digital' Vacuum Microelectronics: Carbon-Nanotube-Based Inverse Majority Gates for High Temperature Applications", Institute of Electrical and Electronic Engineers #978-1-4244-7099-0/10, pp. 203–204, 2010

G.G. Linceo, "*Istoria e Dimonstrazioni Intorno Alle Macchie Solari* (Letters on Sunspots)", Appresso Giacomo Masaerdi, Roma, 1613

F.J. Dyson, "Search for Artificial Stellar Sources of Infrared Radiation", *Science*, **131** (3414), pp. 1667–1668, 3 June 1960

G. Kopp and J.L. Lean, "A new, lower value of total solar irradiance: Evidence and climate significance", *Geophys. Res. Lett.*, **38**, L01706, 2011. doi:10.1029/2010GL045777

Mark Oxborrow, Jonathan D. Breeze, Neil M. Alford, "Room-temperature solid-state maser", *Nature*, **488**, pp. 353–356, 16 August 2012. doi:10.1038/nature11339

G.A. Landis, "Microwave Pushed Interstellar Sail: Starwisp Revisited", AIAA-2000-3337, AIAA 36th Joint Propulsion Conference and Exhibit, Huntsville, Alabama, 17–19 Jul 2000

E. Hecht and A. Zajac, "Optics", Addison-Wesley, Reading, Massachusetts, pp. 352, 1979

B. Khayatian, Y. Rahmat-Samii, R. Porgorzelski, "An Antenna Concept Integrated with Future Solar Sails", *IEEE Antennas and Propagation Society International Symposium*, vol. 2, pp. 742–745, Boston, Mass., July 8–13, 2001

NASA, Press Release 12-157: "NASA Survey Counts Potentially Hazardous Asteroids", 16 May 2012.

Y. Tsuda, O. Mori, R. Funase, H. Sawada, T. Yamamoto, T. Saiki, T. Endo, K. Yonekura, H. Hoshino, J. Kawaguchi, "Achievement of IKAROS—Japanese Deep Space Solar Sail Demonstration Mission", in G. Genta (ed.), *Proceedings of the 7th Symposium of the International Academy of Astronautics*, Aosta, Italy, 10–13 Jul 2011, pp. 79–84, 2011

G.L. Matloff, "Deflecting Earth Threatening Asteroids Using the Solar Collector: An Improved Model", in G. Genta (ed.), pp. 55–60, *op.cit.*

M. Vasile, "Engineers set their sights on asteroid deflection", Univ. Strathclyde, Glasgow, Scotland, 25 Mar 2012; http://www.strath.ac.uk/press/newsreleases/headline_602313_en.html accessed 23 Sep 2012

J.S. Lewis, "Mining the Sky", Basic Books, New York, 1996

Yu. A. Izrael, G.L. Lioznov, A.A. Rasnovski, "Astronautics Potentials and Limits in Solving Problem of Adaptation to the Climate Change", in Yu. A. Izrael, *et al*, (eds.), pp. 174–175, *op. cit.*

M.J. Fogg, "Terraforming: Engineering Planetary Environments", Society of Automotive Engineers, 1995

E. Mallove and G.L. Matloff, "The Starflight Handbook", Wiley, 1989

G.L. Matloff, L. Johnson, and C. Bangs, "Living Off the Land in Space: Green Roads to the Cosmos", Springer/Copernicus Books, 2007

A. Demayo, "Elements in the Earth's Crust", in R.C. Weast, *et al* (eds.), *CRC Handbook of Chemistry and Physics*, 67th edition, Chemical Rubber Press, Boca Raton, Florida, p. F-137, 1986

H.S. Hudson, "A Space Parasol as a Countermeasure Against the Greenhouse Effect", *JBIS*, **44**, pp. 139–41, 1991

L.D. Kunsman and C.L. Carlson, "6.4 Nonferrous Metals", in E. Avallone and T. Baumeister, (eds.), *Mark's Standard Handbook for Mechanical Engineers*, 9th edition, McGraw-Hill, New York, 1987

F. Kreith, "Principles of Sustainable Energy Systems", 2nd ed., CRC Press, Boca Raton, Florida, 2013

Military Advisory Board, "National Security and the Threat of Climate Change", Center for Naval Analysis, Alexandria, Virginia, 2007

The Authors

These four authors have written the preceding two appendices of this volume. This chapter is representative of presentations of research that influences national and supra-national agencies in their efforts to deal with national and global issues such as climate change.

Robert G. Kennedy (PE), a senior systems engineer specializing in green energy at local, regional and national levels, was educated in the classics and foreign languages and has earned a master's degree in national security studies. He invented the Tetrageneration™ concept, worked in robotics at Douglas Aircraft, investigated artificial intelligence at ORNL, and founded Ultramax, a Russian-American trading company located in Oak Ridge,Tennessee. Robert also spent a year as an ASME Congressional Fellow with the House Subcommittee on Space. He has published papers on geo-engineering, terraforming and Cold War History in JBIS, Spaceflight, Smithsonian Air & Space, and Acta Astronautica and has addressed the Russian Academy of Sciences.

Kenneth Roy (PE), an engineer based in Oak Ridge, Tennessee, has earned degrees from the Illinois Institute of Technology and the University of Tenessee. His work on kinetic weapons has been featured in the *Proceedings of the U.S. Naval Institute*. Ken is the inventor of the "Shell World" concept; his papers have been published in *JBIS* and *Acta Astronautica*.

Eric Hughes, a mathematician with a degree from UC Berkeley, is interested in the intersection of technology and society. He is a founder of Cypherpunks, an activist group for privacy and cryptography concerned with the relationship between large-scale power and private life. Eric's research includes equipment that blends modern electronics such as 3D printing with traditional handicraft techniques such as glassblowing.

Dr. David E. Fields, an experimental solid-state physicist, directs the Tamke-Allan Observatory, an optical and radio-astronomy observatory at Rome State

G. Matloff et al., *Harvesting Space for a Greener Earth*,
DOI 10.1007/978-1-4614-9426-3, © Springer Science+Business Media New York 2014

Community College. At Oak Ridge National Laboratory (ORNL), he has specialized on environmental transport and human risk from chemicals and radionuclides. He has also participated in environmental studies in Brazil and Germany. He has consulted for NASA-MSFC on the design of astronaut radiation shields. David has 172 publications, 2 patents and about 100 conference presentations. Current research interests include robot antennas, software-defined radio, radio interferometry and radio-astronomy research.

Index

G. Matloff et al., *Harvesting Space for a Greener Earth*,
DOI 10.1007/978-1-4614-9426-3, © Springer Science+Business Media New York 2014